Plant Management and Soil Improvement in Specialty Crop Production

Plant Management and Soil Improvement in Specialty Crop Production

Guest Editors

Xunfeng Chen
Linchuan Fang

MDPI

Basel • Beijing • Wuhan • Barcelona • Belgrade • Novi Sad • Cluj • Manchester

Guest Editors
Xunfeng Chen
School of Environment and Safety Engineering
Jiangsu University
Zhenjiang
China

Linchuan Fang
Key Laboratory of Green Utilization of Critical Non-metallic Mineral Resources
Wuhan University of Technology
Wuhan
China

Editorial Office
MDPI AG
Grosspeteranlage 5
4052 Basel, Switzerland

This is a reprint of the Special Issue, published open access by the journal *Applied Sciences* (ISSN 2076-3417), freely accessible at: https://www.mdpi.com/journal/applsci/special_issues/4N53FTPBDT.

For citation purposes, cite each article independently as indicated on the article page online and as indicated below:

Lastname, A.A.; Lastname, B.B. Article Title. *Journal Name* **Year**, *Volume Number*, Page Range.

ISBN 978-3-7258-4907-9 (Hbk)
ISBN 978-3-7258-4908-6 (PDF)
https://doi.org/10.3390/books978-3-7258-4908-6

Contents

About the Editors

Xunfeng Chen

Xunfeng Chen is an Assistant Professor and Master Supervisor at the School of Environment and Safety Engineering, Jiangsu University, China. His research focuses on soil remediation, nanotechnology in environmental protection, agricultural pollution control, and plant osmotic stress. Dr. Chen has authored over 30 SCI-indexed publications in journals like the Journal of Hazardous Materials, and holds 7 Chinese patents. Dr. Chen has been awarded several prestigious talent programs, including the Jiangsu Provincial Innovation and Entrepreneurship Doctor ("Double Innovation" Doctor) Program. He serves as Guest Editor for Applied Sciences.

Linchuan Fang

Linchuan Fang is a Professor and Doctoral Supervisor at the School of Resources and Environmental Engineering, Wuhan University of Technology, China. His research focuses on soil heavy-metal pollution control, environmental functional materials, and soil remediation technology. He leads major projects including the National Natural Science Foundation of China (NSFC) grants, and participates in the National Key Research and Development Program of China. Professor Fang has published over 120 academic papers. He serves as Associate Editor of Plant–Environment Interactions, and an Editorial Board Member of Frontiers in Microbiology.

applied sciences

MDPI

Editorial

Plant Management and Soil Improvement in Specialty Crop Production

Xunfeng Chen [1,2,3,*] and Linchuan Fang [4,*]

1 Biofuels Institute, School of Environment and Safety Engineering, Jiangsu University, Zhenjiang 212013, China
2 School of Emergency Management, Jiangsu University, Zhenjiang 212013, China
3 Engineering Research Center of Groundwater Pollution Control and Remediation, Ministry of Education of China, Beijing Normal University, Beijing 100875, China
4 Key Laboratory of Green Utilization of Critical Non-Metallic Mineral Resources, Ministry of Education, Wuhan University of Technology, Wuhan 430070, China
* Correspondence: chenxunfeng1991@163.com (X.C.); flinc629@hotmail.com (L.F.)

Specialty crops, which include fruits, vegetables, nuts, and ornamental plants, play a crucial role in global agriculture and nutrition [1]. Unlike commodity crops, specialty crops often require more intensive management practices to optimize quality, yield, and economic viability. Central to the success of specialty crop production are effective plant management strategies and sustainable soil improvement techniques. As the demand for specialty crops continues to rise, the need for sustainable and efficient plant management practices becomes increasingly crucial [2]. In this editorial, we explore the current practices, challenges, and innovations in these areas, aiming to highlight the importance of integrated approaches for enhancing productivity while minimizing environmental impact.

1. Importance of Plant Management in Specialty Crop Production

Effective plant management encompasses a range of practices aimed at maximizing crop health, yield, and quality. In specialty crop production, where market demands often emphasize superior taste, appearance, and nutritional value, precise management strategies are essential. Integrated pest management (IPM) is a cornerstone of sustainable specialty crop production. By combining biological, cultural, physical, and chemical control tactics, IPM minimizes the use of synthetic pesticides while effectively managing pests and diseases. Techniques such as crop rotation, beneficial insect releases, and monitoring systems are integral to successful IPM programs [3]. Meanwhile, precision agriculture technologies, including GPS-guided machinery, remote sensing, and data analytics, offer precise insights into soil variability, crop health, and water management. These technologies enable farmers to optimize input use, reduce resource wastage, and enhance overall productivity [4]. Also, efficient water management is critical for specialty crop production, particularly in regions prone to water scarcity. Techniques such as drip irrigation, soil moisture sensors, and deficit irrigation strategies help conserve water while maintaining optimal crop growth and yield [5]. It is worth noting that specialty crops often have specific nutrient requirements for optimal growth and quality. Precision fertilization based on soil testing, tissue analysis, and nutrient modeling ensures that crops receive adequate nutrients without excess, resulting in reduced environmental impacts such as nutrient leaching and runoff [6].

Despite advancements, several challenges persist in plant management for specialty crops. Such specialty crops are susceptible to a wide range of pests and diseases, necessitating vigilant monitoring and rapid response strategies. Many specialty crops require labor-intensive management practices, which can be costly and challenging in regions with labor shortages. At the same time, high temperatures, overuse of antibiotics, altered precipitation patterns, and extreme weather events pose risks to crop production, necessitating adaptive management practices [7]. Fortunately, recent innovations are transforming plant management into specialty crop production. Advances in biological control agents, such as predatory insects and microbial biopesticides, offer effective alternatives to conventional pesticides [8]. Breeding for disease resistance, drought tolerance, and improved nutritional profiles through genetic technologies is enhancing crop resilience and quality. Moreover, robotics and automation technologies are revolutionizing labor-intensive tasks such as harvesting and precision application of inputs, resulting in reduced costs and improved efficiency [9].

2. Soil-Improvement Strategies

Healthy soils are fundamental to sustainable specialty crop production, providing essential nutrients, water, and support for plant growth. Soil-improvement strategies aim to enhance soil health, fertility, and resilience to environmental stresses. Cover crops improve soil structure, suppress weeds, and enhance nutrient cycling through biomass incorporation [10]. Increasing soil organic matter content through composting, green manures, and crop residue management enhances soil fertility, water retention, and biological activity [11]. Adoption of soil health assessment tools and indicators helps farmers evaluate soil quality and make informed management decisions [12]. In addition, the use of nanomaterials for soil improvement and remediation is also a very effective and promising approach [13,14]. Reduced tillage and conservation tillage practices minimize soil disturbance, erosion, and carbon loss, preserving soil structure and biodiversity [15].

However, many challenges in soil improvement for specialty corps still remain. The process of balancing nutrient inputs to meet crop demand without causing environmental harm requires precise management and monitoring. Intensive rain events and sloped terrain can lead to soil erosion, necessitating erosion control measures. Heavy machinery and traffic can compact soils, reducing water infiltration and root growth. Increasing prevalence of nanoplastics could negatively impact agricultural production and soil health [16]. Meanwhile, innovative approaches are addressing these challenges and advancing soil improvement techniques. Biochar, a stable form of carbon produced from biomass, improves soil fertility, water holding capacity, and nutrient retention, which can be beneficial to soil improvement [17]. Application of beneficial microbes enhances nutrient availability, disease suppression, and overall soil health [18,19]. In addition, use of gypsum, lime, and other soil amendments corrects pH imbalances and enhances nutrient availability for specialty crops [20].

Effective plant management and soil improvement are critical for sustainable specialty crop production. Advances in integrated pest management, precision agriculture, and soil health enhancement are enhancing productivity while reducing environmental impacts. However, challenges such as pest pressures, labor intensity, and climate change require ongoing innovation and adaptation. By embracing technological advancements, biological control methods, and sustainable practices, growers can optimize crop quality and yield while safeguarding soil health for future generations. Going forward, ongoing new improvements in plant management and soil improvement will lead to more opportunities for accelerated development of specialty crop production.

Funding: This work was supported by the National Agricultural Science and Technology Major Project of China (No. NK202214030304), the National Natural Science Foundation of China (No. 62101216), and the Key Research and Development Program of Zhenjiang (No. SH2023105), and the Open Project Program of Engineering Research Center of Groundwater Pollution Control and Remediation, Ministry of Education of China (GW202406), and Special Research Project of Emergency Management School, Jiangsu University, China (KY-D-19).

Conflicts of Interest: The authors declare no conflicts of interest.

References

1. Vuppalapati, C. *Specialty Crops, in Specialty Crops for Climate Change Adaptation: Strategies for Enhanced Food Security by Using Machine Learning and Artificial Intelligence*; Vuppalapati, C., Ed.; Springer Nature: Cham, Switzerland, 2023; pp. 35–197.
2. Zhao, S.; Yue, C. Risk preferences of commodity crop producers and specialty crop producers: An application of prospect theory. *Agric. Econ.* **2020**, *51*, 359–372. [CrossRef]

3. Abrol, D.P. *Integrated Pest Management: Current Concepts and Ecological Perspective*; Academic Press: Cambridge, MA, USA, 2013.
4. Karunathilake, E.M.B.M.; Le, A.T.; Heo, S.; Chung, Y.S.; Mansoor, S. The path to smart farming: Innovations and opportunities in precision agriculture. *Agriculture* **2023**, *13*, 1593. [CrossRef]
5. Baligar, V.; Fageria, N.; He, Z. Nutrient use efficiency in plants. *Commun. Soil Sci. Plant Anal.* **2001**, *32*, 921–950. [CrossRef]
6. Williams, Z.F. *Evaluation of Sustainable Fertilizers on Soil Health and Yield of Specialty Crops Grown on the Delmarva Peninsula*; University of Maryland Eastern Shore: Princess Anne, MD, USA, 2021.
7. Hu, X.; Zhou, Q.; Luo, Y. Occurrence and source analysis of typical veterinary antibiotics in manure, soil, vegetables and groundwater from organic vegetable bases, northern China. *Environ. Pollut.* **2010**, *158*, 2992–2998. [CrossRef] [PubMed]
8. Dara, S.K. The new integrated pest management paradigm for the modern age. *J. Integr. Pest Manag.* **2019**, *10*, 12. [CrossRef]
9. Mahmud, M.S.A.; Abidin, M.S.Z.; Emmanuel, A.A.; Hasan, H.S. Robotics and automation in agriculture: Present and future applications. *Appl. Model. Simul.* **2020**, *4*, 130–140.
10. Wood, S.A.; Bowman, M. Large-scale farmer-led experiment demonstrates positive impact of cover crops on multiple soil health indicators. *Nat. Food* **2021**, *2*, 97–103. [CrossRef] [PubMed]
11. Shinde, R.; Shahi, D.K.; Mahapatra, P.; Singh, C.S.; Naik, S.K.; Thombare, N.; Singh, A.K. Management of crop residues with special reference to the on-farm utilization methods: A review. *Ind. Crops Prod.* **2022**, *181*, 114772. [CrossRef]
12. Guo, M. Soil health assessment and management: Recent development in science and practices. *Soil Syst.* **2021**, *5*, 61. [CrossRef]
13. Chen, X.; Chu, S.; Chi, Y.; Wang, J.; Wang, R.; You, Y.; Hayat, K.; Khalid, M.; Zhang, D.; Zhou, P.; et al. Unraveling the role of multi-walled carbon nanotubes in a corn-soil system: Plant growth, oxidative stress and heavy metal (loid) s behavior. *Plant Physiol. Biochem.* **2023**, *200*, 107802. [CrossRef] [PubMed]
14. Chen, X.; Wang, J.; Wang, R.; Zhang, D.; Chu, S.; Yang, X.; Hayat, K.; Fan, Z.; Cao, X.; Ok, Y.S.; et al. Insights into growth-promoting effect of nanomaterials: Using transcriptomics and metabolomics to reveal the molecular mechanisms of MWCNTs in enhancing hyperaccumulator under heavy metal (loid) s stress. *J. Hazard. Mater.* **2022**, *439*, 129640. [CrossRef] [PubMed]
15. Kopittke, P.M.; Menzies, N.W.; Wang, P.; McKenna, B.A.; Lombi, E. Soil and the intensification of agriculture for global food security. *Environ. Int.* **2019**, *132*, 105078. [CrossRef] [PubMed]
16. Shi, R.; Liu, W.; Lian, Y.; Wang, X.; Men, S.; Zeb, A.; Wang, Q.; Wang, J.; Li, J.; Zheng, Z.; et al. Toxicity mechanisms of nanoplastics on crop growth, interference of phyllosphere microbes, and evidence for foliar penetration and translocation. *Environ. Sci. Technol.* **2023**, *58*, 1010–1021. [CrossRef] [PubMed]
17. Brtnicky, M.; Datta, R.; Holatko, J.; Bielska, L.; Gusiatin, Z.M.; Kucerik, J.; Hammerschmiedt, T.; Danish, S.; Radziemska, M.; Mravcova, L.; et al. A critical review of the possible adverse effects of biochar in the soil environment. *Sci. Total Environ.* **2021**, *796*, 148756. [CrossRef] [PubMed]
18. Jansson, J.K.; Hofmockel, K.S. Soil microbiomes and climate change. *Nat. Rev. Microbiol.* **2020**, *18*, 35–46. [CrossRef] [PubMed]
19. Naz, M.; Dai, Z.; Hussain, S.; Tariq, M.; Danish, S.; Khan, I.U.; Qi, S.; Du, D. The soil pH and heavy metals revealed their impact on soil microbial community. *J. Environ. Manag.* **2022**, *321*, 115770. [CrossRef]
20. Bamdad, H.; Papari, S.; Lazarovits, G.; Berruti, F. Soil amendments for sustainable agriculture: Microbial organic fertilizers. *Soil Use Manag.* **2022**, *38*, 94–120. [CrossRef]

applied sciences

Article

Effect of Planting *Portulaca oleracea* L. on Improvement of Salt-Affected Soils

Jing Dong [1], Jincheng Xing [1,2,*], Tingting He [1,3], Sunan He [1], Chong Liu [1], Xiaomei Zhu [1], Guoli Sun [1], Kai Wang [1], Lizhou Hong [1] and Zhenhua Zhang [1,3,4,*]

1 Jiangsu Coastal Area Institute of Agricultural Sciences, Yancheng 224002, China; dongjingyc@163.com (J.D.)
2 College of Life Sciences, Anhui Normal University, Wuhu 241004, China
3 Key Laboratory of Saline-Alkali Soil Improvement and Utilization (Coastal Saline-Alkali Lands), Ministry of Agriculture and Rural Affairs, Nanjing 210015, China
4 The School of Agriculture and Environment, The University of Western Australia, Crawley, WA 6009, Australia
* Correspondence: sdauxxx@163.com (J.X.); zhenhua.zhang@uwa.edu.au (Z.Z.)

Abstract

Saline–alkali land is a critical factor limiting agricultural production and ecological restoration. Utilizing salt-tolerant plants for bioremediation represents an environmentally friendly and sustainable approach to soil management. This study employed the highly salt-tolerant crop *Portulaca oleracea* L. cv. "Su Ma Chi Xian 3" as the test material. A plot experiment was established in coastal saline soils with planting *P. a- oleracea* (P) and no planting (CK) under three blocks with the different salt levels (S1: 2.16 g/kg; S2: 4.08 g/kg; S3: 5.43 g/kg) to systematically evaluate its salt accumulation capacity and effects on soil physicochemical properties. The results demonstrated that *P. oleracea* exhibited adaptability across all three salinity levels, with aboveground biomass following the trend PS2 > PS3 > PS1. The ash salt contents removed through harvesting were 1.29, 2.03, and 1.74 t/ha, respectively, in PS1, PS2, and PS3. Compared to no planting, a significant reduction in bulk density was observed in the 0–10 and 10–20 cm soil layers ($p < 0.05$). A significant increase in porosity by 9.72%, 16.29%, and 12.61% was found under PS1, PS2, and PS3, respectively, in the 0–10 cm soil layer. Soil salinity decreased by 34.20%, 50.23%, and 48.26%, in the 0–10 cm soil layer and by 14.43%, 32.30%, and 26.42% in the 10–20 cm soil layer under PS1, PS2, and PS3, respectively. The pH exhibited a significant reduction under the planting treatment in the 0–10 cm layer. A significant increase in organic matter content by 13.70%, 12.44%, and 13.55%, under PS1, PS2, and PS3, respectively, was observed in the 0–10 cm soil layer. The activities of invertase and urease were significantly enhanced in the 0–10 and 10–20 cm soil layers, and the activity of alkaline phosphatase also exhibited a significant increase in the 0–10 cm layer under the planting treatment. This study indicated that cultivating *P. oleracea* could effectively facilitate the improvement of coastal saline soils by optimizing soil structure, reducing salinity, increasing organic matter, and activating the soil enzyme system, thereby providing theoretical and technical foundations for ecological restoration and sustainable agricultural utilization of saline–alkali lands.

Keywords: halophytic plants; phytoremediation; soil reclamation; saline–alkali land

1. Introduction

Soil salinization has emerged as a globally escalating environmental crisis that threatens agricultural sustainability and food security [1]. Current estimates indicate that approximately 1 billion hectares of land worldwide are affected by salinization, with this figure continuing to rise at an alarming rate [2]. In recent decades, the rapid advancement of urbanization has led to a large amount of cultivated land becoming occupied, resulting in continuous reduction in cultivated land area, which undoubtedly brings huge pressure to food production [3]. For China, a country with scarce land resources, a small cultivated land area, and a wide distribution of salt-affected soil, saline–alkali land management is an important measure to ensure national food security from a strategic height and an important link to alleviate the contradiction between people and land and promote the sound and rapid development of the national economy [4,5]. Conventional amelioration approaches, including hydraulic engineering measures and chemical amendments, have demonstrated limited effectiveness in addressing salinization challenges [6–8]. These traditional methods are frequently constrained by high implementation costs, operational complexity, and the potential for secondary environmental contamination. In contrast, the utilization of salt-tolerant plants for phytoremediation represents a promising alternative strategy. This phytoremediation approach offers distinct advantages through its cost-effectiveness, environmental compatibility, and potential economic returns from cultivated species. As such, it embodies a sustainable, eco-friendly, and economically viable solution for saline–alkali soil reclamation and can be widely applied worldwide for the restoration of such degraded lands [9].

Studies have demonstrated that certain halophytic species can absorb saline ions from salt-affected soils and enhance both the physicochemical properties and structural characteristics of saline soils [10,11]. Purslane, *Portulaca oleracea* L, is an herbaceous plant belonging to the Portulacaceae family that exhibits strong stress resistance (including tolerance to salinity, drought, and poor soil conditions), high photosynthetic efficiency, minimal susceptibility to pests and diseases, and low requirements for pesticides and fertilizers [12,13]. It is widely distributed across the globe. Under suitable temperature and light conditions, *P. oleracea* develops a unique high-yield growth pattern due to its C4 photosynthetic advantage, strong regenerative capacity, and efficient resource utilization [14]. With a short growth cycle, *P. oleracea* can trigger an axillary bud regeneration mechanism by retaining 2–3 basal nodes during harvesting, resulting in multiple regenerative branches and allowing for repeated harvesting.

In recent years, studies have revealed that *P. oleracea* possesses significant potential for the remediation of saline–alkali soils. *P. oleracea* demonstrates the ability to grow normally under certain levels of salt stress, exhibiting strong salt tolerance [15,16]. Remarkably, it can absorb salts from the soil and thrive and complete its life cycle even in high salt-affected conditions [17]. Nevertheless, the studies on *P. oleracea* have primarily focused on its physiological traits, with limited exploration of its role as a dominant species for ecological restoration in salt–alkali land. The effects of planting *P oleracea* in the ecological remediation of coastal salt-affected soils remain poorly understood. Therefore, the objectives of this study were to (1) determine *P. oleracea* capacity of salt removal in coastal saline soils with different salinity concentrations, (2) compare its effects on the physicochemical properties of salt-affected soils, and (3) further clarify its phytoremediation potential in saline–alkali lands. The findings are expected to provide a scientific basis for the application of purslane in saline–alkali soil remediation and establish a theoretical foundation for the biological improvement of saline–alkali soils.

2. Materials and Methods

2.1. Characteristics of the Experimental Site

The experimental site is located at the tidal flat experimental base of Jinhai Farm in Dafeng District, Yancheng City, Jiangsu Province, China (32°59′42″ N, 120°49′32″ E), approximately 6 km east of the Yellow Sea. The site lies within the northern subtropical monsoon climate zone, with an annual precipitation of 900–1300 mm, predominantly concentrated between June and August. Three field blocks with different salinity levels (S1: 2.16‰; S2: 4.08‰; S3: 5.43‰) were selected as experimental sites. The soil used in the experiment is classified as alluvial saline soil. The basic properties of three experimental blocks in 0–20 cm soil are presented in Table 1.

Table 1. Basic properties of three blocks with different salt levels in 0–20 cm soil.

Blocks	Soil Bulk Density (g/cm^3)	Soil Salinity (g/kg)	pH	Soil Organic Matter (g/kg)	Available Nitrogen (mg/kg)	Available Phosphorus (mg/kg)	Available Potassium (mg/kg)
S1	1.47 ± 0.03	2.16 ± 0.08	8.01 ± 0.01	8.34 ± 0.03	28.59 ± 0.80	10.42 ± 0.27	195.53 ± 4.92
S2	1.51 ± 0.02	4.08 ± 0.07	7.95 ± 0.02	6.93 ± 0.04	26.45 ± 0.81	9.23 ± 0.29	187.67 ± 9.51
S3	1.48 ± 0.03	5.43 ± 0.09	8.05 ± 0.02	6.64 ± 0.06	24.72 ± 0.75	8.92 ± 0.26	176.6 ± 7.41

Note: The data is presented as mean ± SD (n = 3).

2.2. Experimental Design and Setup

The experiment was conducted from April to August 2024. The six treatments were two planting treatments (planting *P. oleracea* [(P) and non-planting (CK)] × three salinity levels (S1, S2, and S3). Each treatment was replicated three times. A total of 18 experimental plots (4 m × 10 m) were established. The *P. oleracea* variety "Su Ma Chi Xian 3" was chosen as the test material, which was bred by Jiangsu Coastal Area Institute of Agricultural Sciences, Yancheng, China. The seeding rate is 15 kg/ha. No fertilization was applied during the experiment. The experiment commenced on 15 April 2024, with two cuttings of *P. oleracea* taken on 13 June and 13 July, retaining approximately 8 cm of the rootstock each time. The final harvest was conducted on 16 August 2024. Soil samples were collected after the harvest of *P. oleracea*.

2.3. Measurement Methods

2.3.1. Plant Biomass

The fresh and dry biomass of *P. oleracea* aboveground parts was measured after each cutting event and at the final harvest to estimate the total aboveground biomass. Fresh plant materials were weighed, then plant samples of about 2 kg were taken in triplicate to measure the moisture content of the plants. The plant samples were deactivated at 105 °C for 0.5 h and dried at 80 °C for more than 48 h until a constant weight was achieved to determine the dry weight. The dry weight was calculated according to fresh weight time moisture content.

2.3.2. Ash Content in Plant and Salt Accumulation by Plants

The plant materials were ashed in a muffle furnace at 550 °C for 8 h. After cooling, the weight of the ash was measured to determine the ash content. We used biomass of above-ground plant time ash content in aboveground plants to calculate the salt accumulation capacity of plants based on Wang et al. [18]. The salt accumulation capacity of *P. oleracea* was calculated using the following formula:

$$\text{Total amount of salt accumulation (kg/ha)} = \text{Total aboveground biomass (kg/ ha)} \times \text{Ash content (\%)} \quad (1)$$

2.3.3. Soil Physicochemical and Biological Analysis

The salt content of the soil profile from 0 to 60 cm was determined, with sampling depths at 0–10, 10–20, 20–40, and 40–60 cm. The total soil salt content was determined by the weight loss of a 5:1 (water soil mass ratio) extract after being oven-dried at 105 °C to a constant weight [19].

The measurements of soil bulk density, porosity, pH, organic matter content (SOM), and enzyme activity were conducted across three soil layers (0–10, 10–20, and 20–40 cm). The soil bulk density and porosity were measured by inserting steel cylinders. Soil pH was measured by drying and grinding the soil samples, followed by extraction with a 5:1 (water–soil mass ratio) solution and then determining the pH of the extract using a pH meter (Sartorius, PB-10, Gottingen, Germany). The soil organic matter content was determined using a potassium dichromate oxidation technique combined with external heating [20].

Soil invertase activity was determined using 3, 5-dinitrosalicylic acid colorimetry, soil urease activity was measured by the phenol–sodium colorimetric method, and soil alkaline phosphatase activity was calculated by colorimetry using p-nitrophenyl phosphate disodium as the substrate [21,22].

We measured soil physicochemical and biologic properties across three soil layers (0–10, 10–20, and 20–40 cm), except for salt content, which was measured across four layers (0–10, 10–20, 20–40, and 40–60 cm). The objective of this study was mainly focused on phytoremediation of salt from soils. The salt in the soils is very easy to move up and down with water, but the other parameters are relatively stable and less changed at 40–60 cm soil depth, since *P. oleracea* is a shallow-rooted plant.

2.4. Statistical Analysis

A one-way analysis of variance (ANOVA) was performed to analyze the significant difference of biomass, ash content in plants, and salt accumulation by plants among treatments, and a two-way ANOVA was conducted to analyze significant changes of soil physiochemical and biological properties among treatments through SPSS Statistics 20.0 software. Duncan's test was used to test significant differences between treatments when ANOVA showed significant effects ($p < 0.05$). Graphs were created using Excel 2016.

3. Results

3.1. Aboveground Biomass of *P. oleracea*

The highest fresh weight of aboveground part among treatments was observed in PS2 treatment, which significantly increased by 56.75% and 15.15%, respectively, compared to PS1 and PS3. The aboveground dry weight of *P. oleracea* under the three salt levels (PS1, PS2, and PS3) reached 5.00 t/ha, 7.61 t/ha, and 6.54 t/ha, respectively. Notably, the aboveground biomass of *P. oleracea* was the highest in PS2, exceeding that of PS1 and PS3 by 52.20% and 16.36%, respectively (Table 2).

Table 2. Aboveground biomass of *P. oleracea* in the salt-affected soils.

Treatment	Fresh Weight (t/ha)				Dry Weight (t/ha)			
	First Harvest	Second Harvest	Third Harvest	Total	First Harvest	Second Harvest	Third Harvest	Total
PS1	11.54 ± 0.20c	14.78 ± 0.21c	30.42 ± 0.41c	56.74 ± 0.43c	1.02 ± 0.02c	1.30 ± 0.02c	2.68 ± 0.04c	5.00 ± 0.04c
PS2	17.44 ± 0.13a	23.71 ± 0.36a	47.79 ± 0.38a	88.94 ± 0.73a	1.49 ± 0.01a	2.03 ± 0.03a	4.09 ± 0.03a	7.61 ± 0.06a
PS3	14.48 ± 0.28b	20.07 ± 0.47b	42.69 ± 0.31b	77.24 ± 0.47b	1.23 ± 0.02b	1.70 ± 0.04b	3.62 ± 0.03b	6.54 ± 0.04b

Note: Means ± SD ($n = 3$) with different letters within columns are significantly different ($p < 0.05$).

3.2. Accumulation of Salt by *P. oleracea*

The effect of planting *P. oleracea* on salt accumulation is shown in Table 3. There was no significant difference in ash content in plants among the treatments. The salt accumulation in aboveground *P. oleracea* was significantly influenced by soil salinity levels. The salt accumulation reached 1.29, 2.03, and 1.74 t/ha, respectively, in PS1, PS2, and PS3 treatments.

Table 3. Ash content in plants and salt accumulation by *P. oleracea* in the salt-affected soils.

Treatment	Ash Content (%)			Salt Accumulation (t/ha)			
	First Harvest	Second Harvest	Third Harvest	First Harvest	Second Harvest	Third Harvest	Total
PS1	26.68 ± 0.53a	25.81 ± 0.48a	25.55 ± 0.56a	0.27 ± 0.01c	0.34 ± 0.01c	0.69 ± 0.02c	1.29 ± 0.02c
PS2	27.03 ± 0.31a	26.35 ± 0.34a	26.57 ± 0.77a	0.40 ± 0.01a	0.53 ± 0.01a	1.09 ± 0.04a	2.03 ± 0.05a
PS3	26.90 ± 0.39a	26.25 ± 0.46a	26.51 ± 0.50a	0.33 ± 0.01b	0.45 ± 0.02b	0.96 ± 0.02b	1.74 ± 0.01b

Note: Means ± SD (n = 3) with different letters within columns are significantly different ($p < 0.05$).

3.3. *Effect of Planting P. oleracea on Soil Physiochemical Properties in Salt-Affected Soils*

3.3.1. Salinity Distribution in Soils

The salt content was significantly lower in planting soils than non-planting soils at 0–10 and 10–20 cm soil layers but not significantly different at 20–40 and 40–60 cm (Figure 1). Compared to non-planting, the salt contents significantly decreased by 34.20%, 50.23%, and 48.26% in the 0–10 cm soil layer and by 14.43%, 32.30%, and 26.42% in the 10–20 cm layer under PS1, PS2, and PS3 treatments, respectively. The differences in salt content gradually decreased with the increase in soil depth between the planting and non-planting soils.

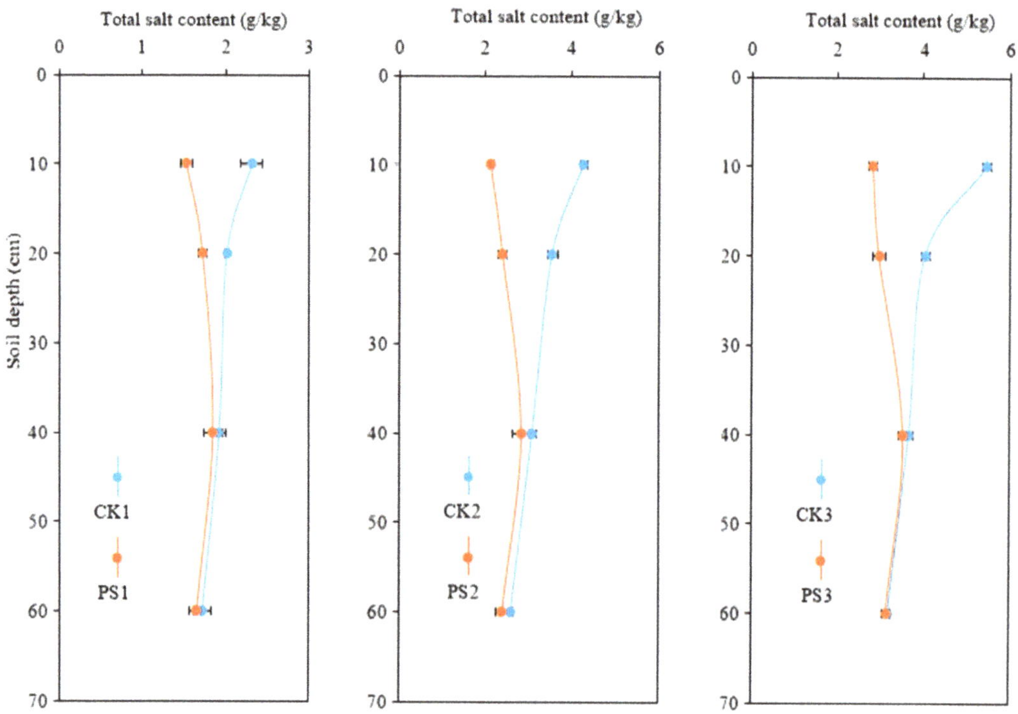

Figure 1. Total salt contents in soil under the planting (PS) and no-planting (CK) treatments at different soil depths.

3.3.2. Soil Bulk Density

The soil bulk density in the 0–10 and 10–20 cm layers was significantly higher in planting soils than non-planting soils ($p < 0.05$) (Figure 2). Compared to CK, the bulk density in the 0–10 cm and 10–20 cm soil layers significantly decreased by 8.05% and 7.05% in PS1, by 12.81% and 7.97% in PS2, and by 10.27% and 7.16% in PS3, respectively.

The soil bulk density in the 20–40 cm layer decreased to different degrees in the planting treatments, but the difference was not significant.

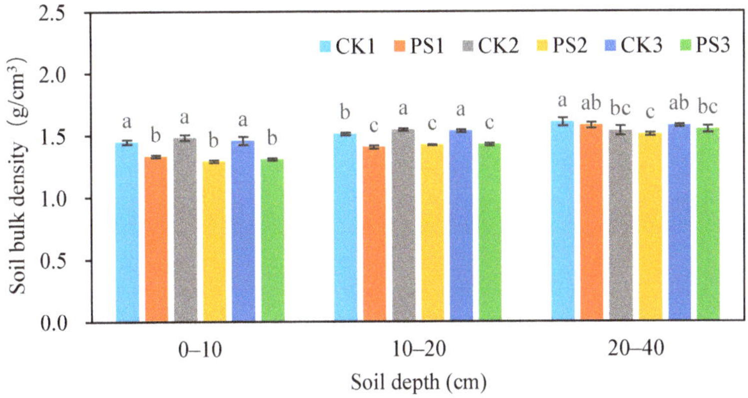

Figure 2. Effect of planting *P. oleracea* on soil bulk density in salt-affected soils. Different lowercase letters above bars at the same soil depth indicate a significant difference at $p < 0.05$.

3.3.3. Soil Porosity

Compared to non-planting soil, the soil porosity of all layers increased to varying degrees after planting *P. oleracea* (Figure 3). The soil porosity was significant higher in planting soils than non-planting soils in the 0–10 and 10–20 cm soil layers ($p < 0.05$) but not in the 20–40 cm soil layer. The soil porosity at 0–10 cm increased by 9.72%, 16.29%, and 12.61%, respectively, in PS1, PS2, and PS3.

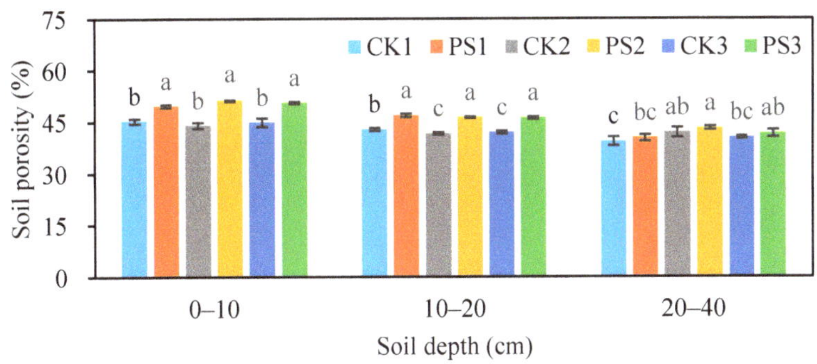

Figure 3. Effect of planting *P. oleracea* on soil porosity in salt-affected soils. Different lowercase letters above bars at the same soil depth indicate a significant difference at $p < 0.05$.

3.3.4. Soil pH

The soil pH at 0–10 cm was significantly lower in planting soil than non-planting soil, but there was no significant difference in pH value in the 10–20 and 20–40 cm soil layers ($p > 0.05$) (Figure 4). Compared to non-planting, the pH value at 0–10 cm decreased by 1.96%, 1.81%, and 1.22%, respectively, in the planting soils with three salt levels.

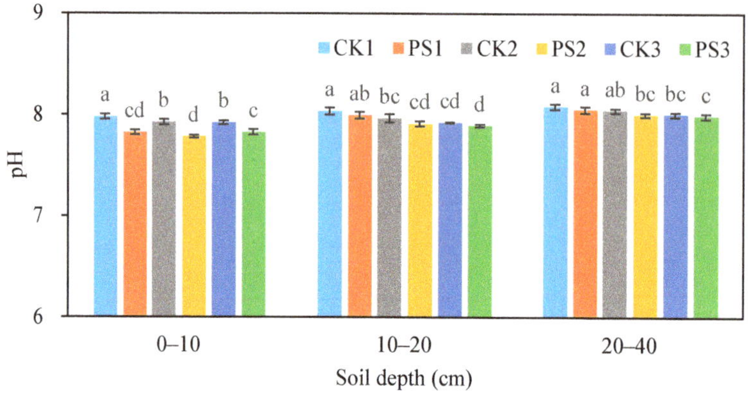

Figure 4. Effect of planting *P. oleracea* on soil pH in salt-affected soils. Different lowercase letters above bars at the same soil depth indicate a significant difference at $p < 0.05$.

3.3.5. Soil Organic Matter

The organic matter contents were significantly higher in planting than non-planting soils at 0–10 and 10–20 cm but not at 20–40 cm (Figure 5). Compared to non-planting, the organic matter contents in the 0–10 cm soil layer significantly increased by 13.75%, 12.34%, and 13.56%, respectively, in PS1, PS2, and PS3, while they increased by 4.59%, 6.90%, and 7.92% in the 10–20 cm soil layer.

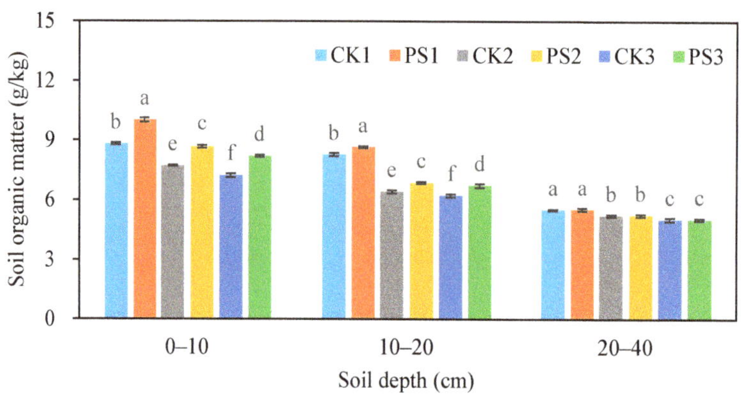

Figure 5. Effect of planting *P. oleracea* on soil organic matter contents in salt-affected soils. Different lowercase letters above bars at the same soil depth indicate a significant difference at $p < 0.05$.

3.4. Effect of Planting *P. oleracea* on Soil Enzyme Activity

3.4.1. Soil Invertase Activity

The soil invertase activity showed a significant increase in planting treatments in the 0–10 and 10-20 cm soil layers, but no significant difference was observed in the 20–40 cm layer (Figure 6). Compared to non-planting, the invertase activity in the 0–10 cm soil layer exhibited 2.22-, 2.11-, and 1.92-fold increases in three salt-level planting treatments, while demonstrating comparatively smaller increases at 10–20 cm.

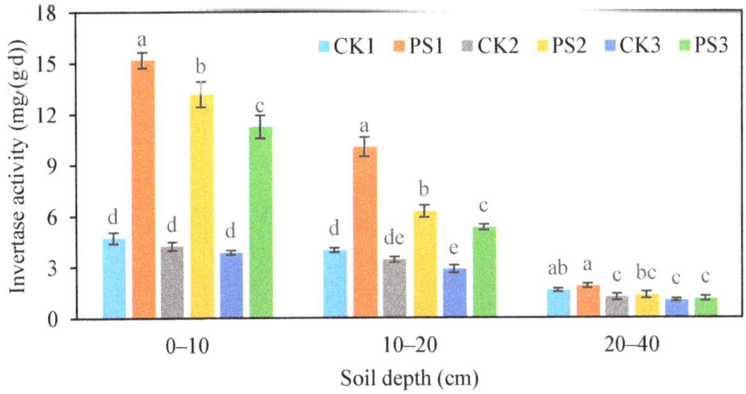

Figure 6. Effect of planting *P. oleracea* on invertase activity in salt-affected soils. Different lowercase letters above bars at the same soil depth indicate a significant difference at $p < 0.05$.

3.4.2. Soil Urease Activity

The soil urease activity significantly increased in planting treatments in the 0–10 and 10–20 cm soil layers, while there was no significant difference in the 20–40 cm soil layer (Figure 7). Compared to non-planting, the activity of urease in the 0–10 cm soil layer increased by 56.25%, 61.36%, and 47.62%, respectively, in PS1, PS2, and PS3, and by 55.56%, 50.00%, and 41.38%, respectively, in the 10–20 cm soil layer.

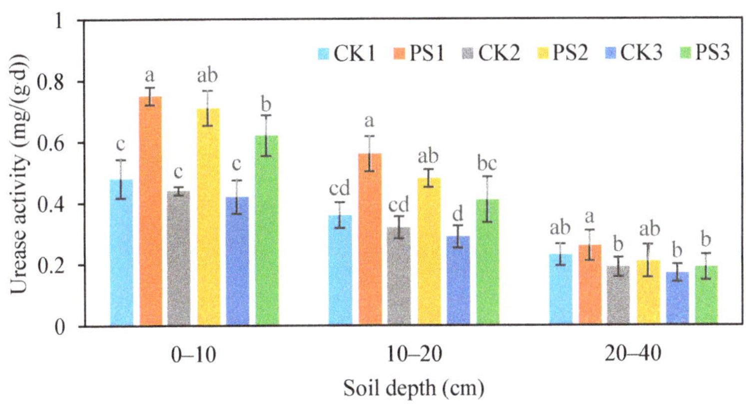

Figure 7. Effect of planting *P. oleracea* on urease activity in salt-affected soils. Different lowercase letters above bars at the same soil depth indicate a significant difference at $p < 0.05$.

3.4.3. Soil Alkaline Phosphatase Activity

The soil alkaline phosphatase activity significantly increased in planting treatments in the 0–10 cm soil layer, while there was no significant difference in the 10–20 and 20–40 cm soil layers (Figure 8). Compared to non-planting, the activity of alkaline phosphatase in the 0–10 cm soil layer increased by 13.88%, 10.00%, and 17.03%, respectively, in PS1, PS2, and PS3.

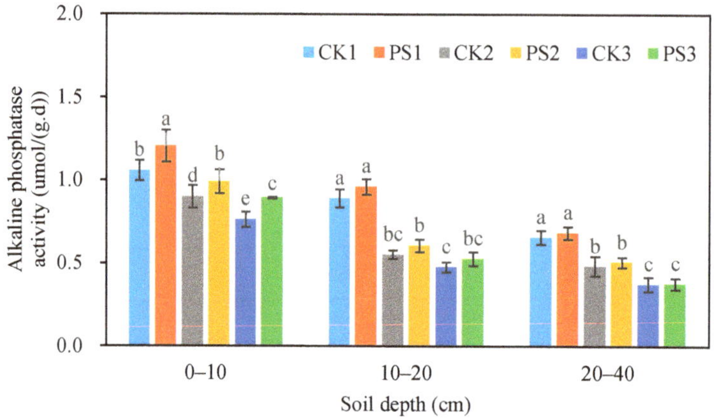

Figure 8. Effect of planting *P. oleracea* on alkaline phosphatase activity in salt-affected soils. Different lowercase letters above bars at the same soil depth indicate a significant difference at $p < 0.05$.

4. Discussion

4.1. Phytoremediation by P. oleracea in Salt-Affected Soils

In the phytoremediation of salt-affected soil, the efficiency of salt extraction highly depends on the characteristics of the selected halophyte species [23]. Ideal remediation plants should possess high biomass yield and strong salt absorption capacity. Therefore, when selecting remediation species, priority should be given to halophytes of the salt-accumulating type to fully utilize their efficient salt absorption and accumulation capabilities [24]. This study further revealed the unique ecological functions of *P. oleracea* in saline environments, particularly its ability to accumulate salt in its aboveground parts. It was found that the trend of ash salt content accumulated by *P. oleracea* in the different treatments was PS2 > PS3 > PS1. When harvested, the salt removal reached 1.29, 2.03, and 1.74 t/ha, respectively, in planting treatments under three salinity levels (Table 3). Notably, the ash salt accumulation (2.03 t/ha) in the PS2 treatment was particularly significant, equivalent to half of the ash salt accumulation of *Suaeda salsa*. Meanwhile, the ash salt accumulation in the PS3 treatment (1.74 t/ha) was comparable to that of *S. salsa* at the seedling stage [18]. A similar study reported by Kilic et al. indicated that *P. oleracea* can effectively absorb salt in the soil, which can effectively remove 65 kg/ha of Na^+ and 210 kg/ha of Cl^- in a growing season and accumulate these salts in aboveground parts, thereby reducing soil salt content [17]. These results indicate that *P. oleracea* exhibits similar salt-tolerance mechanisms to typical halophytes such as *Halogeton glomeratus* and *Suaeda glauca* [25]. Notably, *P. oleracea* not only adapts well to high-salinity environments but also effectively reduces soil salt content by accumulating salt in its aboveground biomass. This characteristic gives it significant application value in salt-affected soil remediation. Through regular cutting, *P. oleracea* can serve as an efficient phytoremediation tool, helping to reduce soil salinity levels while providing a sustainable solution for the ecological restoration of salt affected soil. Additionally, the salt accumulation capacity of *P. oleracea* complements its salt tolerance mechanisms, further highlighting its adaptability and ecological value in adverse environments. Meanwhile, the salinity in the 0–10 and 10–20 cm soil layers decreased after harvest (Figure 1), demonstrating the strong desalination capacity of *P. oleracea*. The possible mechanisms of desalination included the following: (1) the absorption and translocation of salt from the soil to the aboveground parts through transpiration, ultimately removing salt through harvesting; (2) the secretion of organic acids and other substances by *P. oleracea* roots, which promote the dissolution and leaching of salt ions in the soil; (3) the promotion of soil aggregate formation by root exudates, reducing surface salt accumulation.

4.2. Planting *P. oleracea* Improves Soil Physiochemical and Biological Properties in Salt-Affected Soils

Planting *P. oleracea* could improve soil structure in salt-affected soil. Soil bulk density is an important physical indicator reflecting soil compactness and structural condition, often referred to as the physical structural fertility of soil [26]. Excessive bulk density indicates reduced soil porosity, which may lead to excessive water retention, thereby affecting the growth and development of plant roots. In this study, after planting *P. oleracea*, the bulk density significantly decreased (Figure 2), and porosity increased in the 0–10 and 10–20 cm soil layers (Figure 3), indicating that the root growth of *P. oleracea* effectively improved soil structure and enhanced soil aeration and water permeability. Studies have demonstrated that cultivating salt-tolerant plants can mitigate salt buildup in soils by enhancing their structural properties. For instance, Araya et al. found that halophyte cultivation reduced soil bulk density while enhancing porosity and hydraulic conductivity, facilitating the leaching of salts from saline–alkali soils [27]. Similarly, Qadir et al. observed that the plants improved soil physicochemical properties, promoting the removal of Na^+ and soluble salts through enhanced leaching [28].

Planting *P. oleracea* could increase soil fertility in salt-affected soil. In the present study, the significant decrease in pH in the 0–10 cm soil layer (Figure 4) could be related to the secretion of organic acids by roots and the acidic substances produced by the decomposition of soil organic matter [29,30]. The reduction in pH helps improve the availability of nutrients such as phosphorus, iron, and zinc, promoting plant growth [31]. Additionally, the organic matter content in the 0–10 cm soil layer increased (Figure 5), indicating that *P. oleracea* primarily increases surface soil organic matter content through aboveground litter and root exudates. The increase in organic matter content helps improve soil structure, enhance water and nutrient retention, and promote soil microbial activity, thereby improving soil fertility and promoting the health of the soil ecosystem [31]. Soil enzyme activity is an important indicator reflecting soil fertility and ecological functions [32]. In this study, the activities of invertase and urease in the 0–20 cm soil layer significantly increased (Figures 6 and 7), indicating that the cultivation of *P. oleracea* promoted soil carbon and nitrogen cycling. The enhancement of invertase activity facilitated the decomposition of carbohydrates in the soil, providing an energy source for soil microorganisms. The increase in urease activity, on the other hand, promotes the decomposition of nitrogen-containing organic compounds such as urea, releasing ammonium nitrogen that is readily absorbable by plants [33]. Additionally, the activity of alkaline phosphatase in the 0–10 cm soil layer showed a significant increase (Figure 8), suggesting that the cultivation of *P. oleracea* enhanced the mineralization of organic phosphorus in the soil, thereby improving the availability of soil phosphorus [34]. The soil microbial community composition possibly changed under the ameliorative effect of various halophytes or salt-tolerant plants on saline soils [25]. Hence, it is necessary to conduct metagenomic analysis to reveal the phytoremediation effects of halophytes or salt-tolerant plants in saline soils in future studies.

5. Conclusions

This study demonstrated that planting *P. oleracea* significantly reduced the pH values and soluble salt contents in salt-affected soil while improving soil structure and increasing soil organic matter contents and enzyme activities. These multifunctional characteristics highlighted the significant application value of *P. oleracea* for salt-affected soil improvement.

Future studies should further elucidate the molecular mechanisms underlying the salt tolerance and remediation functions of *P. oleracea*. Additionally, optimizing agronomic management practices (such as cutting frequency and fertilizer management) could enhance its ecological remediation efficiency, thereby promoting the large-scale application of this technology in the management of coastal saline–alkali lands.

Author Contributions: Conceptualization, J.D. and J.X.; methodology, J.D.; formal analysis, J.D. and T.H.; investigation, S.H., C.L. and X.Z.; resources, J.X.; data curation, G.S., K.W. and L.H.; writing—original draft preparation, J.D.; writing—review and editing, J.D., J.X. and Z.Z.; project administration, J.X. and Z.Z.; supervision, Z.Z.; funding acquisition, J.D., J.X. and Z.Z. All authors have read and agreed to the published version of the manuscript.

Funding: This research was funded by Jiangsu Provincial Crop Germplasm Resource Bank (Salt-tolerant plants) of Yancheng Municipal Bureau of Agriculture and Rural Affairs (JS-ZW-K12) and the Scientific and Technological Innovation Fund of Carbon Emissions Peak and Neutrality of Jiangsu Provincial Department of Science and Technology (BE2022304).

Institutional Review Board Statement: Not applicable.

Informed Consent Statement: Not applicable.

Data Availability Statement: The datasets used and analyzed during the current study are available from the corresponding author upon reasonable request.

Acknowledgments: We thank all the participants involved in this study.

Conflicts of Interest: The authors declare no conflicts of interest.

References

1. Sinha, S.; Ibha, S.; Vaibhav, S.; Singh, C.P.; Pratap, S.R.; Vishal, P. Potential risk assessment of soil salinity to agroecosystem sustainability: Current status and management strategies. *Sci. Total Environ.* **2021**, *764*, 144164.
2. Zhang, H.; Yu, F.; Xie, P.; Sun, S.; Qiao, X.; Tang, S.; Chen, C.; Yang, S.; Mei, C.; Yang, D.; et al. A Gγ protein regulates alkaline sensitivity in crops. *Science* **2023**, *379*, 6638. [CrossRef] [PubMed]
3. Bai, X.; Shi, P.; Liu, Y. Society: Realizing China's urban dream. *Nature* **2014**, *509*, 158–160. [CrossRef] [PubMed]
4. Cao, X.; Sun, B.; Chen, H.; Zhou, J.; Song, X.; Liu, X.; Deng, X.; Li, X.; Zhao, Y.; Zhang, J.; et al. Approaches and research progresses of marginal land productivity expansion and ecological benefit improvement in China. *Bull. Chin. Acad. Sci.* **2022**, *36*, 336–348.
5. Du, Y.; Liu, X.; Zhang, L.; Zhou, W. Drip irrigation in agricultural saline-alkali land controls soil salinity and improves crop yield: Evidence from a global meta-analysis. *Sci. Total Environ.* **2023**, *880*, 163226. [CrossRef]
6. Heng, T.; He, X.; Yang, L.; Xu, X.; Feng, Y. Mechanism of Saline–Alkali land improvement using subsurface pipe and vertical well drainage measures and its response to agricultural soil ecosystem. *Environ. Pollut.* **2022**, *293*, 118583. [CrossRef]
7. Liu, M.; Wang, C.; Liu, X.; Lu, Y.; Wang, Y. Saline-alkali soil applied with vermicompost and humic acid fertilizer improved macroaggregate microstructure to enhance salt leaching and inhibit nitrogen losses. *Appl. Soil Ecol.* **2020**, *156*, 103705. [CrossRef]
8. Zhao, Y.; Li, Y.; Wang, S.; Wang, J.; Xu, L. Combined application of a straw layer and flue gas desulphurization gypsum to reduce soil salinity and alkalinity. *Pedosphere* **2020**, *30*, 226–235. [CrossRef]
9. Chu, L.; Yuan, S.; Chen, D.; Kang, Y.; Shaghaleh, H.; Okla, M.K.; AbdElgawad, H.; Hamoud, Y.A. Changes in salinity and vegetation growth under different land use types during the reclamation in coastal saline soil. *Chemosphere* **2024**, *366*, 143427. [CrossRef]
10. Ma, P.; Shi, Z.; Diao, F.; Hao, L.; Zhang, J.; Xu, J.; Wang, L.; Dang, Z.; Guo, W. Effects of arbuscular mycorrhizal fungi on growth and Na^+ accumulation of *Suaeda glauca* (Bunge) grown in salinized wetland soils. *Appl. Soil Ecol.* **2021**, *166*, 104065. [CrossRef]
11. Lv, S.; Jiang, P.; Tai, F.; Wang, D.; Feng, J.; Fan, P.; Bao, H.; Li, Y. The V-ATPase subunit A is essential for salt tolerance through participating in vacuolar Na^+ compartmentalization in *Salicornia europaea*. *Planta* **2017**, *246*, 1177–1187. [CrossRef] [PubMed]
12. Wang, X.; Ma, X.; Yan, G.; Hua, L.; Liu, H.; Huang, W.; Liang, Z.; Chao, Q.; Hibberd, J.M.; Jiao, Y.; et al. Gene duplications facilitate C4-CAM compatibility in common purslane. *Plant physiol.* **2023**, *193*, 2622–2639. [CrossRef] [PubMed]
13. Jin, R.; Wang, Y.; Liu, R.; Gou, J.; Chan, Z. Physiological and metabolic changes of purslane (*Portulaca oleracea* L.) in response to drought, heat and combined stresses. *Front. Plant Sci.* **2016**, *6*, 1123. [CrossRef] [PubMed]
14. MorenoVillena, J.J.; Zhou, H.; Gilman, I.S.; Tausta, S.L.; Cheung, C.Y.M.; Edwards, E.J. Spatial resolution of an integrated C_4+CAM photosynthetic metabolism. *Sci. Adv.* **2022**, *8*, eabn2349. [CrossRef]

15. Yazdani, B.R.; Karimi, M.; Soltangheisi, A. Purslane (*Portulaca oleracea* L.) salt tolerance assessment. *Soil Sci. Plant Nutr.* **2023**, *69*, 250–259. [CrossRef]
16. Mohamed, M.H.M.; Ali, M.M.E.; Zewail, R.M.Y.; Liava, V.; Petropoulos, S.A. Response of purslane plants grown under salinity stress and biostimulant formulations. *Plants* **2024**, *13*, 2431. [CrossRef]
17. Kiliç, C.C.; Kukul, S.Y.; Anaç, D. Performance of purslane (*Portulaca oleracea* L.) as a salt-removing crop. *Agric. Water Manag.* **2008**, *95*, 854–858. [CrossRef]
18. Wang, L.; Wang, X.; Jiang, L.; Zhang, K.; Mohsin, T.; Tian, C.; Zhao, Z. Reclamation of saline soil by planting annual euhalophyte *Suaeda salsa* with drip irrigation: A three-year field experiment in arid northwestern China. *Ecol. Eng.* **2020**, *159*, 106090. [CrossRef]
19. Liang, J.; Shi, W. Cotton/halophytes intercropping decreases salt accumulation and improves soil physicochemical properties and crop productivity in saline-alkali soils under mulched drip irrigation: A three-year field experiment. *Field Crops Res.* **2021**, *262*, 108027. [CrossRef]
20. Walkley, A.; Black, I.A. An examination of the degtjareff method for determining soil organic matter, and a proposed modification of the chromic acid titration method. *Soil Sci.* **1934**, *37*, 29–38. [CrossRef]
21. Zhu, Y.; Zhong, M.; Li, W.; Qiu, Y.; Wang, H.; Lv, X. Cotton straw biochar and Bacillus compound biofertilizer decreased Cd migration in alkaline soil: Insights from relationship between soil key metabolites and key bacteria. *Ecotoxicol. Environ. Saf.* **2022**, *232*, 113293. [CrossRef]
22. Bandara, T.; Herath, I.; Kumarathilaka, P.; Hseu, Z.Y.; Ok, Y.S.; Vithanage, M. Efficacy of woody biomass and biochar for alleviating heavy metal bioavailability in serpentine soil. *Environ. Geochem. Health* **2017**, *39*, 391–401. [CrossRef] [PubMed]
23. Fatemeh, A.; Nayer, M.; Moslem, S. Halophytes play important role in phytoremediation of salt-affected soils in the bed of Urmia Lake, Iran. *Sci. Rep.* **2022**, *12*, 12223.
24. Hasan, H.; Shloul, T.; Alomari, B.; Alhadidi, L.; Mazahreh, N. Phytoremediation ability of *Panicum maximum* and *Salicornia europaea* irrigated with treated wastewater for salt elements in the soil. *J. Saudi Soc. Agric. Sci.* **2024**, *23*, 451–457. [CrossRef]
25. Wang, J.; Song, M.; Yao, L.; Li, P.; Si, E.; Li, B.; Meng, Y.; Ma, X.; Yang, K.; Zhang, H.; et al. Metagenomic analysis reveal the phytoremediation effects of monocropping and intercropping of halophytes *Halogeton glomeratus* and *Suaeda glauca* in saline soil of Northwestern China. *BMC Plant Biol.* **2025**, *25*, 213. [CrossRef] [PubMed]
26. Panagos, P.; Rosa, D.D.; Liakos, L.; Labouyrie, M.; Borrelli, P.; Ballabio, C. Soil bulk density assessment in Europe. *Agric. Ecosyst. Environ.* **2024**, *364*, 108907. [CrossRef]
27. Araya, T.; Mlahlwa, A.V.; Elbasit, M.A.M.A.; Newete, S.W. The impact of *Tamarix invasion* on the soil physicochemical properties. *Sci. Rep.* **2022**, *12*, 5750. [CrossRef]
28. Qadir, M.; Steffens, D.; Yan, F.; Schubert, S. Sodium removal from a calcareous saline–sodic soil through leaching and plant uptake during phytoremediation. *Land Degrad. Dev.* **2003**, *14*, 301–307. [CrossRef]
29. Zhang, J.; Chen, G.; Li, Y.; Zhang, J.; Zhong, L.; Li, L.; Zhong, S.; Gu, R. Phlomoides rotata adapts to low-nitrogen environments by promoting root growth and increasing root organic acid exudate. *BMC Plant Biol.* **2024**, *24*, 1234. [CrossRef]
30. Clarholm, M.; Skyllberg, U.; Rosling, A. Organic acid induced release of nutrients from metal-stabilized soil organic matter-The unbutton model. *Soil Biol. Biochem.* **2015**, *84*, 168–176. [CrossRef]
31. Xiao, M.; Jiang, S.; Li, J.; Li, W.; Fu, P.; Liu, G.; Chen, J. Synergistic effects of bio-organic fertilizer and different soil amendments on salt reduction, soil fertility, and yield enhancement in salt-affected coastal soils. *Soil Tillage Res.* **2025**, *248*, 106433. [CrossRef]
32. Wang, L.; Hamel, C.; Lu, P.; Wang, J.; Sun, D.; Wang, Y.; Lee, S.; Gan, G.Y. Using enzyme activities as an indicator of soil fertility in grassland-an academic dilemma. *Front. Plant Sci.* **2023**, *14*, 1175946. [CrossRef] [PubMed]
33. Krajewska, B. Urease-aided calcium carbonate mineralization for engineering applications: A review. *J. Adv. Res.* **2017**, *13*, 59–67. [CrossRef]
34. Turner, B.L.; Haygarth, P.M. Phosphatase activity in temperate pasture soils: Potential regulation of labile organic phosphorus turnover by phosphodiesterase activity. *Sci. Total Environ.* **2005**, *344*, 27–36. [CrossRef] [PubMed]

applied sciences

MDPI

Article

Investigating the Effects of Fish Effluents as Organic Fertilisers on Basil (*Ocimum basilicum*)

Lorenzo Fruscella [1], Benz Kotzen [1], Marcos Paradelo Perez [2] and Sarah Milliken [1,*]

[1] School of Design, University of Greenwich, Park Row, London SE10 9LS, UK; b.kotzen@greenwich.ac.uk (B.K.)

[2] Natural Resources Institute, University of Greenwich, Central Avenue, Chatham Maritime, Kent ME4 4TB, UK; m.paradeloperez@greenwich.ac.uk

* Correspondence: s.milliken@greenwich.ac.uk

Abstract: Whilst the potential of fish effluents as nutrient sources for crop production has been demonstrated, their use in the European Union remains prohibited in organic farming. In this study, we investigate the efficacy in greenhouse basil cultivation of two types of fish effluents (filtered 'fish water' and unfiltered 'fish sludge') from an aquaponic system, and assess their role in maintaining and enhancing soil fertility as well as their potential to create a 'living soil', which are two of the prerequisites for organic certification in the EU. To evaluate the contribution of fish effluents to plant growth in comparison with soil nutrients, basil plants were grown in pots containing two types of substrate: compost-free (without organic matter) and with compost (with organic matter). The results indicate that fish water and fish sludge demonstrate significant potential as fertilisers and outperform compost in certain parameters, such as plant biomass. The results also align with existing literature by demonstrating the positive impact of compost on soil microbial diversity, underscoring its role in fostering plant health. Although the treatments did not show differences in microbial composition at the genus level, the higher microbial diversity observed following fish effluent application highlights its potential for promoting 'living soil'. This research underscores the need for continued exploration of the implications of compost application in conjunction with fish effluent fertilisation on soil microbial communities and the production of specialty crops such as herbs.

Keywords: basil; fish; effluents; aquaculture; aquaponics; organic; sludge

1. Introduction

Whilst the potential of wastes from aquaculture farms, termed fish effluents, as a nutrient source for crop production has been demonstrated, their use is not sanctioned in organic farming within the prevailing European regulatory framework—Regulation (EU) 2018/848 [1], which lays down the rules for organic production, and Regulation (EU) 2023/121 [2], which specifies the types of fertilisers that are authorised. However, using fish wastewater as a fertiliser can benefit the environment in several ways. By diverting nutrient-rich effluent from aquatic ecosystems, it can help mitigate eutrophication caused by excess nitrogen and phosphorus [3]. Moreover, repurposing fish sludge in agriculture reduces the carbon footprint associated with manufacturing and transporting synthetic fertilisers [4]. Economically, fish wastewater can be cost-effective for growers located near aquaculture facilities, as it lowers reliance on commercial fertilisers and cuts waste disposal costs [5]. Fish manure (sludge) has a chemical composition akin to livestock manure [6], and fish effluents (sludge or fish tank water) have been used successfully to fertilise various food crops, including tomato [7–13], pepper [14], chicory [15], lettuce [16–19], cucumber [20], potato, soybean, and onion [21]. Up to 50% (in dry matter) of the feed consumed by fish is expelled as solids, forming sludge [22], and a significant portion of the nutrients introduced into an aquaculture system is wasted through the routine discharge of this nutrient-rich matter, which is primarily composed of fish excrement and unconsumed feed [23]. However, in most aquaculture farms, the sludge is released as sewage, and only occasionally is it dried and repurposed as fertiliser [24]. Aquaculture sludge holds considerable promise for enhancing plant growth due to its nutrient content [25], and a substantial amount of the essential macro- and micronutrients can be derived from this solid waste, which has been found to contain 7–32% of the total nitrogen and 30–84% of the total phosphorus present in wastewater [26]. The sludge is particularly abundant in phosphorus, a crucial macronutrient for plant growth which has been recognised as a Critical Raw Material in the EU due to its high supply risk, and conventional filtration in aquaponic systems eliminates over 80% of it [27]. The nutrient composition of filtered aquaculture effluents differs from that of sludge, with the latter offering a much richer and broader nutritional spectrum. Indeed, in comparison to sludge, filtered aquaculture effluent typically lacks phosphorus, potassium, calcium, and micronutrients, notably iron, molybdenum, and manganese [10].

Should the fish feed be organic in provenance, then the application of fish effluents could supply crops with organic fertiliser through a natural process observed in aquatic environments, where fish effluents contribute to plant growth in ponds and lakes [28]. Additionally, recycling wastewater that would otherwise pollute water bodies could offer a potential remedy for the environmental challenges associated with the eutrophication of aquatic ecosystems [29]. In this manner, the use of fish effluents could also fulfil an environmental conservation role by partially replacing the need for synthetic fertilisers, the production of which consumes significant energy [30], often sourced in environmentally damaging ways, such as phosphorus mining [31,32].

In this study, the efficacy of two types of fish effluents (filtered and unfiltered) from an aquaponic system was tested to grow basil in greenhouse conditions. The main objectives were to test their viability as a source of fertiliser for the greenhouse cultivation of herbs, and to investigate their potential for maintaining and enhancing soil fertility and for enhancing the production of 'living soil', which are conditions that need to be fulfilled for organic certification in Europe. In order to ascertain the role of fish effluents in the growth of the plants as opposed to the nutrients already present in soil, the basil plants were grown in pots filled with two types of substrate, one compost-free and the other with compost. The filtered fish water is hereafter referred to as 'fish water', the unfiltered fish water as 'fish sludge' and the water from the tap as 'tap water'. When referring to both fish water and fish sludge, they are referred to as 'fish effluents'. A key novelty of this study is its direct comparison of compost and fish effluents as fertilisers for the cultivation of basil in greenhouse conditions. By evaluating plant growth and soil microbial properties across treatments, this work provides a unique perspective on the relative advantages of compost versus fish effluents, advancing our understanding of how these fertiliser strategies can be optimised for both high-value crop production and soil health in sustainable farming systems.

2. Materials and Methods

2.1. Outline

The experiments took place inside a 7.2 × 3.4 m (24 m^2) greenhouse located on the second floor of the School of Design of the University of Greenwich in London, UK. The experiment started on July 13th and ended on 26 August 2021, lasting a total of 44 days. Basil plants were grown in plastic pots (one plant per pot) placed within plastic trays under grow lights. Two types of potting substrates, one with compost and one without, and two fish effluent types (fish water and fish sludge) were used in combination.

Tap water was sourced from the University's potable water system, whilst fish effluents were obtained from the University of Greenwich aquaponics greenhouse. The greenhouse accommodated a floating raft aquaponic system stocked with approximately 300 Nile tilapia (*Oreochromis niloticus*), alongside a variety of plant species such as beans, squash, mint, tomato, cucumber, okra, fern, turmeric, cape gooseberry, melon, and various ornamentals. The Nile tilapia, weighing between 100 and 500 g, were fed twice daily, once in the morning and once in the afternoon, with Aller Aqua Primo 6 mm sinking pellet feed (37% Crude Protein, 12% Crude Fat, 32.5% Nitrogen-free Extracts, 7% Ash, 3.5% Fibre, 1% Phosphorus, 19.6 MJ Gross Energy, 16 MJ Digestible Energy), totalling 500 g of feed per day.

Most of the fish sludge was removed from the water before it entered the plant (hydroponic) compartments. The first step of filtration involved the clarifier, where the water from the fish tanks was drawn from the bottom, and as it flowed into the clarifier, the settleable solids dropped. To ensure that almost all large solids were removed before the water entered the hydroponic section, single or multiple tanks with orchard netting were employed, achieving almost total removal of solids. The fish sludge was collected from the bottom of the clarifiers, while the fish water was taken from the sump, a compartment where the filtered water flowed and was further aerated before heading to the plant units.

2.2. Crop Choice

F1 hybrid basil *Ocimum basilicum* 'Aroma 2' was used in the experiment, grown from seed. Basil is fast-growing and of high value, and as a result has become one of the main crops grown in aquaponics in both commercial and research settings [33–38]. Basil is also a herb, and herbs constitute an exception to rule 1.1 of Annex II of Regulation (EU) 2018/848 on the mandatory connection with the subsoil and bedrock—this means that basil can be grown in pots and still be certified as organic in the EU. Furthermore, the morphology of the crop allows for easy measurement of growth parameters such as height and stem diameter, the leaves are easily collectable for further analysis, and the crop has been widely studied and analysed for biomass and nutrient content when subjected to a variety of different growing conditions. Finally, the 'Aroma 2' cultivar is one of the most popular for hydroponic growers of basil, chosen for its uniformity, fast growth, classic flavour and aroma profile, as well as resistance to *Fusarium* [39].

2.3. Treatments

Two substrates and three fertilisation regimes were combined for a total of six treatments, each replicated four times: compost-free substrate irrigated with tap water (treatment IT); compost-free substrate irrigated with fish water (treatment IF); compost-free substrate irrigated with fish water and every third day with fish sludge (treatment IFS); compost substrate irrigated with tap water (treatment CT); compost substrate irrigated with fish water (treatment CF), and compost substrate irrigated with fish water and every third day with fish sludge (treatment CFS). The pots were randomly distributed in four blocks. The composition of the tap water was established, and the composition of the two fish effluents was analysed (Table 1). All plants were irrigated from the top and given 10 mL of either tap water or fish effluent per day.

Table 1. Parameters of the tap water, fish water, and fish sludge

Parameter	Tap Water [1]	Fish Water [2]	Fish Sludge [2]
pH	7.60 [3]	6.7	5.8
Calcium carbonate concentration	323 ppm	-	-
Conductivity at 20 °C	594 µS/cm	-	-
Conductivity at 25 °C	-	791 µS/cm	1020 µS/cm
Turbidity	<0.09 FTU	-	-
Ammonium as NH_4	0.12 mg/L	-	-
Nitrate as NO_3	27.2 mg/L	44.3 mg/L	69.6 mg/L
Nitrate/nitrite calculation	0.55 mg/L	-	-
Nitrite as NO_2	0.013 mg/L	-	-
Hardness (total) as $CaCO_3$	244 mg/L	-	-
Iron as Fe	3.6 µg/l	-	-
Magnesium	4.3 mg/L	12.33 mg/L	18.34 mg/L
Alkalinity (HCO_3)	-	36 mg/L	13 mg/L
Sulphate (SO_4)	-	161.8 mg/L	204.4 mg/L
Boron	-	0.01 mg/L	0.03 mg/L
Sodium	-	44.0 mg/L	50.5 mg/L
Chloride	-	59.2 mg/L	69.8 mg/L
Phosphorus as P	-	4.4 mg/L	12.8 mg/L
Potassium	-	1.5 mg/L	3.5 mg/L
Calcium	-	112.6 mg/L	136.4 mg/L
Carbonate	-	<10 mg/L	<10 mg/L
Total dissolved solids	-	553.7 mg/L	714 mg/L

[1] Water parameters for postcode SE10 9BD obtained from https://www.thameswater.co.uk/help/water-quality/check-your-water-quality#/results/SE109BD (accessed on 2 June 2021). [2] Analysed by NRM Labs on 11 June 2021. NRM Labs, Coopers Bridge, Braziers Lane, Winkfield Row, Bracknell RG42 6NS, United Kingdom. [3] pH measured consistently at the University of Greenwich shows a value of approximately 8.40.

2.4. Substrate Mixes

A review of the physical, chemical, and biological properties of soil ecosystems revealed that there appears to be no clear definition of what constitutes soil; this is reflected in Regulation (EU) 2018/848 [1], where no definition of 'soil', nor of what constitutes soil, is given. With regard to the experimental setup, it was considered that a comparison between a substrate mix with compost and one without compost would be useful in order to determine the influence of compost on the substrate that was exposed to fish effluents. It was therefore decided to include both substrate types, one nutrient-rich where compost was included, and one nutrient-poor without compost. To ensure that the substrates were as similar as possible in composition and texture, both substrates were manufactured from raw ingredients using the Royal Horticultural Society guidelines for creating a John Innes Potting Compost [40]. Two substrate types were thus devised, using only materials approved in organic certification: a substrate without compost, having 3 parts loam, 3 parts sand, and 5 parts coir; and a substrate with compost, having 3 parts loam, 3 parts sand, 3.5 parts coir, and 1.5 parts compost (13.6%). The two substrate types were analysed for nutrients and physicochemical parameters (Table 2).

All nutrient values, except for manganese and calcium, were higher in the compost substrate, reflecting the higher nutrient load supplied by the compost.

Table 2. Values for pH, elements, and physical parameters of the two initial substrate types used.

Parameter	Compost-Free Substrate	Compost Substrate
pH	8	7.81
P	18 mg/L	36.80 mg/L
K	218.35 mg/L	538.20 mg/L
Mg	55.90 mg/L	111.65 mg/L
Mn	7.88 mg/L	7.32 mg/L
Cu	1.56 mg/L	1.97 mg/L
B	0.97 mg/L	1.35 mg/L
Na	37.50 mg/L	85 mg/L
Zn	2.03 mg/L	5.40 mg/L
Ca	1835 mg/L	1700 mg/L
Av Fe	11.96 mg/L	23.70 mg/L
Nitrate-N	61.44 kg/ha	361 kg/ha
Ammonium-N	2.80 kg/ha	7.96 kg/ha
Soil mineral nitrogen	64 kg/ha	369 kg/ha
Organic matter LOI [1]	5.44%	6.22%
Av SO_4	47.02 mg/L	56.86 mg/L
Cation exchange capacity	13.60 meq/100 g	14.7 meq/100 g

[1] Loss on ignition.

2.5. Experimental Setup and Growth Conditions

A total of five 8 cm pots were placed in each tray (Figure 1), and the trays (dimensions: 21 × 16 × 5 cm) for each treatment were randomly positioned next to one another in two rows over two metal shelves.

Figure 1. Arrangement of pots inside tray.

Basil seeds were planted in seedling trays filled with John Innes seed substrate no. 1, and then transferred into the experimental pots when they were at a height of 2.5–3 cm, by washing the substrate off the roots of each plant with tap water and planting the bare rooted seedlings in each pot. The greenhouse was not temperature-controlled, but kept warm between 22 and 27 °C. The air temperature was managed by opening the window whenever it became too warm for the plants. The plants were watered in the morning. Three parallel rows of 115 cm long tubular grow lights (power: 35 w; input: AC 220–240 v and 50–60 Hz; manufacturer: Philips, Eindhoven, the Netherlands), three per shelf, were placed directly above each tray at a distance of 35 cm from the top of the trays. The grow lights were set on a 16-hr light/8-hr darkness cycle. All trays were moved counterclockwise by one tray once every four days in order to guarantee uniform light exposure throughout the experiment. All the plants were harvested by hand, and the substrate carefully removed so as not to damage the roots. The roots were washed with tap water to remove the remaining substrate. The leaf weight and whole plant weight measured were fresh weight, and the moisture level was assumed to be the same for all plants.

2.6. Measurements and Analyses

Plant yields were measured, and soil analyses were conducted to reveal the composition of the substrate before and after fertilisation with fish effluents. The effects of fish effluent fertilisation on plant growth, health, and quality were assessed by means of vegetable tissue analysis, growth rate calculations, and the overall health and appearance of the crop. The plant yield measurements were undertaken on all plants, and the weight measurements were performed in batches per replicate. The concentration of nitrogen in leaf samples was measured using a CHN analyser (model: Thermo Scientific FlashEA 1112, Waltham, MA, USA) at the University of Greenwich, Medway Campus. The whole leaf mass from each replicate was harvested, freeze-dried, and powdered using a mortar; three samples of the powdered product from each replicate were then analysed.

The analyses of the macro- and micronutrients in the fish effluent and substrate samples were performed by NRM Labs, the UK's largest independent provider of agronomic and environmental waste analysis for land-based industries, based in Bracknell, UK. The methods are described in Appendix A.

For soil microbial analysis, three initial substrate samples were taken from each substrate type prior to the beginning of the experiment, and four samples per treatment were taken after three weeks, halfway through the experiment, and on the day the experiment ended. The vials were then frozen at −21 °C until shipped for analysis by Biome Makers (Sacramento, California). DNA extraction was performed with the DNeasy PowerLyzer PowerSoil kit from Qiagen, Venlo, the Netherlands. To characterise the bacterial, archaeal, and fungal microbial communities associated with bulk soils and rhizosphere samples, the 16S rRNA and internal transcribed spacer (ITS) marker regions were selected. The 16S primer was used for sequencing bacteria and archaea, and the ITS primer was used for sequencing fungi. Primers were removed from paired-end reads using Cutadapt 3.5, and the trimmed reads were merged with a minimum overlapping of 100 nucleotides. Next, the sequences were quality-filtered by Expected Error with a maximum value of 1.0. Because of their highly conserved length, 16S reads were subjected to an additional filtering step in order to remove extreme sequence lengths. After quality pre-processing, reads with single nucleotide differences were iteratively clustered together to form ASVs (Amplicon Sequencing Variants) using Swarm v3. De novo chimaeras and remaining singletons were subsequently removed. Finally, the ASVs were compared against Biome Makers' proprietary internal reference database of amplicons using a global alignment, with 97% identity to select the best hit; in cases of multiple best hits with identical qualities, the pipeline automatically adjusted the ASV result resolution to the nearest common ancestor of the hits' taxonomies, which can decrease it to genus or family level. The reference database of amplicons was built using internal manually curated taxonomies from the latest version available of SILVA 138.1 for 16S sequences.

2.7. Statistical Analyses

Whenever possible, the data collected were then analysed statistically using R 4.1.0 software [41]; 2-way ANOVA and Tukey HSD tests were used to analyse statistical differences between the treatments. For every value set, the data distribution was explored and histograms generated to identify the presence of any outliers and test for normality. For all the experiments, no random factors were considered, and the fixed factors were the treatments; the response was a function of the treatments. For the plant yield analyses, unless otherwise specified, each plant or plant part was measured, and the average was used in the ANOVA model, instead of using the whole replicate unit as a random factor. The standard errors were calculated from the analysis of variance residuals.

3. Results

3.1. Plant Growth and Yield

All plants grew in height and weight (Figure 2), and only one plant died of unknown causes (in replicate IT2).

Figure 2. Complete display of the two shelves where the plants were grown, showing the individual replicates (black trays), five pots per tray, and grow lights on the last day of the experiment.

Visually, and on measurement, it was evident that the plants from treatment IT—plants grown in compost-free substrate and irrigated with tap water—were much smaller and less developed than the rest, with an average height of 15.06 cm, compared to the other treatments ranging between 23.24 cm and 25.19 cm (Figure 3).

Figure 3. Comparison of randomly selected trays, one from each treatment, showing the clear visual difference of treatment IT, top left photo, from the rest of the treatments, reflecting its significantly lower product yield.

The growth yield values, including % of nitrogen by dry weight, are shown in Table 3, while the values in boxplot form are shown in Figure 4.

Table 3. Mean values and standard deviation values (±) for growth measurements. Different letters indicate statistically significant differences between treatments ($p < 0.05$), using Tukey's Honest Significant Difference (HSD) test.

Treatment	Height (cm)	Stem Diameter (mm)	Leaf Biomass (g)	Nitrogen (%)
IT	15.06 ± 1.75 b	1.84 ± 0.13 b	7.12 ± 0.85 d	2.55 ± 0.18 c
IF	23.85 ± 2.43 a	2.39 ± 0.23 ab	23.75 ± 1.66 c	4.00 ± 0.21 b
IFS	25.19 ± 2.85 a	2.51 ± 0.35 a	28.25 ± 1.66 bc	4.44 ± 0.42 ab
CT	24.58 ± 2.83 a	2.73 ± 0.29 a	23.50 ± 1.69 c	3.01 ± 0.36 c
CF	24.56 ± 1.21 a	2.75 ± 0.36 a	31.62 ± 1.93 ab	4.40 ± 0.15 ab
CFS	23.24 ± 2.42 a	2.64 ± 0.28 a	35.75 ± 3.07 a	4.90 ± 0.26 a

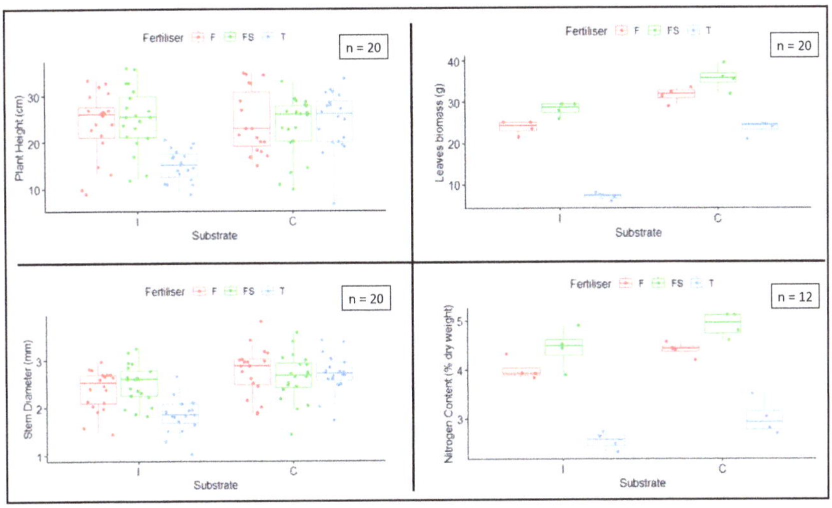

Figure 4. Boxplots for plant height (**top left**), leaf biomass (**top right**), stem diameter (**bottom left**), and nitrogen weight % (**bottom right**), with fertiliser type (F: fish water; FS: fish sludge; T: tap water) and substrate type (I: compost-free substrate; C: compost substrate). Sample sizes (n) in the top right corner of each plot.

In all measurements, the residues of the model were found to be normal, and no outliers were identified. Based on the above results, substrate, fertiliser, and the interaction between substrate and fertiliser all had a significant effect ($p < 0.05$) in all measurements undertaken, except for the nitrogen dry weight percentage, where only substrate and fertiliser had an effect, but not the interaction between them. The fish sludge treatment (FS) outperformed the fish water treatment (F) and the tap water treatment (T) for leaf biomass and nitrogen dry weight percentage in both substrate types (Table 3), with the following pattern: FS > F > T. For the leaf biomass measurement, in fact, and for the compost-free substrate, fish sludge (FS) produced an average biomass of 28.25 g, fish water (F) 23.75 g, and tap water (T) 7.12 g. For the compost substrate, the average leaf biomass was 35.75 g for fish sludge (FS), 31.62 g for fish water (F), and 23.5 g for tap water (T). For the nitrogen dry weight percentage, in the compost-free substrate, fish sludge (FS) produced an average percentage of 4.44, followed by 4.0 for fish water (F), and 2.55 for tap water (T). In the compost substrate treatment, fish sludge (FS) produced a percentage of 4.9, followed by 4.4 for fish water (F), and lastly by 3.01 for tap water (T). The same trend is observed for the stem diameter measurement of the compost-free substrate, where fish sludge (FS) produced an average diameter of 2.64 mm, followed by fish water (F) at 2.75 mm, and tap water (T) at 2.73 mm. However, in the compost substrate the fish water treatment (F) performed better than the fish sludge treatment (FS) for stem diameter (2.75 mm vs. 2.64 mm) and plant height (24.56 mm vs. 23.24 mm). For the leaf biomass measurement, the effects of tap water and compost substrate (CT) were in a very close range to the ones from the compost-free substrate and fish water (IF), at 23.5 g vs. 23.75 average weights. Overall, the fish sludge treatment performed the best. The degrees of freedom (df), F-values, and p-values are reported in Table 4.

Table 4. Two-way ANOVA tables (df, F-values, and p-values) for the measurements, divided by substrate type, fertiliser type, and the interaction between substrate and fertiliser.

Measurement	Substrate/Fertiliser	df	F-Value	p-Value
Plant height	Substrate	1	5.032	0.027
	Fertiliser	2	5.973	0.003
	Substrate/Fertiliser	2	8.820	<0.001
Leaf biomass	Substrate	1	181.56	<0.001
	Fertiliser	2	162.18	<0.001
	Substrate/Fertiliser	2	13.61	<0.001
Stem diameter	Substrate	1	31.870	<0.001
	Fertiliser	2	5.454	0.005
	Substrate/Fertiliser	2	7.458	0.001
Nitrogen weight %	Substrate	1	14.959	0.001
	Fertiliser	2	98.907	<0.001
	Substrate/Fertiliser	2	0.035	0.966

Stem weights were also measured (Figure 5); however, given the lack of sufficient replicates in the measurements (the measurements were taken per treatment batch), complex statistics with reliable results were not achievable.

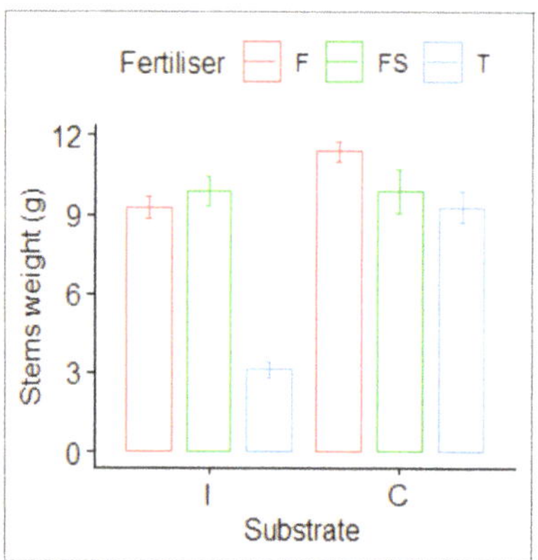

Figure 5. Stem fresh weights, arranged by substrate type (I: compost-free substrate; C: compost substrate) and fertiliser type (F: fish water; FS: fish sludge; T: tap water).

For stem weight and for all fertiliser types, compost substrate (C) performed better than compost free substrate (I), except for the fish sludge treatment (FS), where the values were equal (9.88 g). In both substrate types, tap water (T) produced the lowest yields, at an average of 3.13 g for IT and 9.25 g for CT; however, the highest yield was produced by fish water for compost substrate (CF), with an average weight of 11.38 g, and by fish sludge for compost-free substrate (IFS), with an average weight of 9.88 g.

3.2. Soil Microbiome

In all treatments, the final species count was higher than the initial one; this was also true for the IT treatment, where only tap water was used (Figure 6).

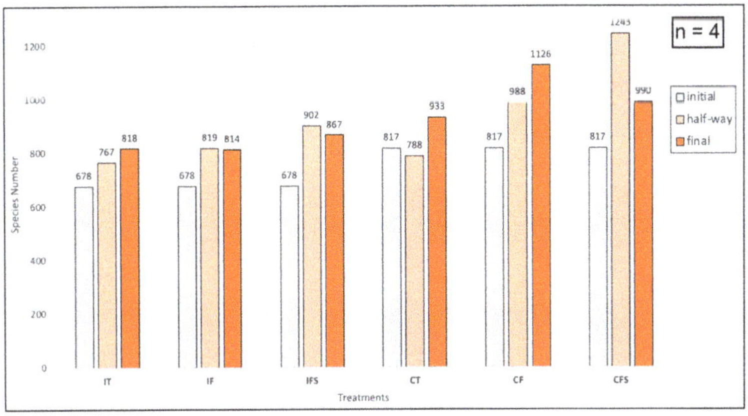

Figure 6. Total species (bacteria, archaea, and fungi) diversity across treatments at initial, halfway, and final experimental stages, with relative species number. The same initial compost-free substrate was used for treatments IT, IF and IFS, and the same initial compost substrate was used for treatments CT, CF, and CFS. Sample size (n) in the top right corner.

In treatments IF, IFS, and CFS, the species count was higher (819, 902, and 1245, respectively) halfway through the experiment than at the end (814, 867, and 990, respectively). The initial compost-free substrate, used in treatments IT, IF, and IFS, was lower in species count than the compost substrate used in treatments CT, CF, and CFS (678 vs. 817). This difference was also reflected in the final species count, given that all final species counts from treatments CT, CF, and CFS (933, 1126, and 990, respectively) were higher than all species counts from treatments IT, IF, and IFS (818, 814, and 867, respectively). The shared species number between the two initial substrates was 372 species.

The most prevalent archaeal and bacterial (Figure 7) and fungal (Figure 8) genera across treatments in halfway and final samples are displayed.

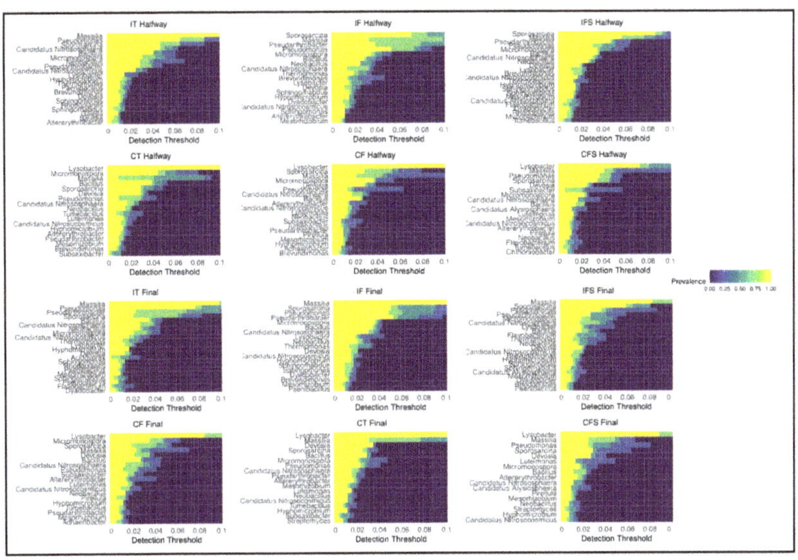

Figure 7. Heatmaps of the most prevalent bacterial and archaeal genera both in the middle (**top two rows**) and at the end (**bottom two rows**) of the experiment, across all treatments.

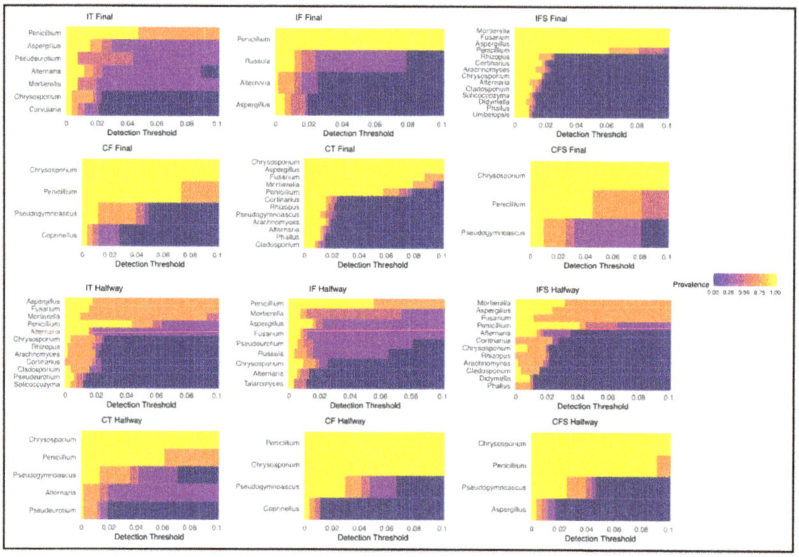

Figure 8. Heatmaps of the most prevalent fungal genera both in the middle (**top two rows**) and at the end (**bottom two rows**) of the experiment, across all treatments.

Regarding bacterial and archaeal genera across treatments, and taking into account the six most prevalent genera, *Massilia* and *Sporosarcina* were present in all treatments, in both halfway and final samples. In halfway samples, *Micromonospora* was present only in treatments with compost-free substrate (IT, IF, and IFS), and *Lysobacter* and *Devosia* were present only in samples with compost substrate (CT, CF, and CFS). In final samples, *Pseudarthrobacter* was only present in treatments with compost-free substrate (IT, IF, and IFS), and *Lysobacter* was only present in samples with compost substrate (CT, CF, and CFS). Regarding fungal genera across treatments, and taking into account the six most prevalent genera, *Penicillium* was present in all treatments, in both halfway and final samples. In halfway samples, *Fusarium* was present only in treatments with compost-free substrate (IT, IF, and IFS), and *Pseudogymnoascus* was present only in samples with compost substrate (CT, CF, and CFS). In both bacterial and archaeal, and fungal analyses, most genera were found across multiple treatments and, in general, little variation was found, including when different fertiliser sources were considered.

4. Discussion

The yield results showed that the presence of compost and fish effluents satisfied the nutrient requirements of the plants, whilst the plants that were grown in a compost-free substrate and watered with tap water (treatment IT) did not have their nutritional requirements met. The results also indicated that fish effluents fully substituted for compost in terms of the nutrition that the plants needed for leaf biomass. The significant effect of combining compost and fertilisation with the fish effluents on the yield of the plants was confirmed by the 2-way ANOVA analysis (Figures 4 and 5). In fact, with the sole exception of nitrogen dry weight percentage, the variables substrate, fertiliser, and the interaction between the two all had a significant effect ($p < 0.05$) in all measurements undertaken. In most cases fish sludge produced the highest yields, reflecting its higher nutrient concentration which was also demonstrated by a previous study on onions [42]. In the case of total fresh weight, compost substrate enhanced the yield, and fish effluents had a positive effect, within each substrate type, with fish water resulting in an almost three-fold increase in weight compared with tap water in compost-free substrate (average weight for IT was 10.25 g vs. 33 g for IF), similar to the influence of compost (average weight for CT was 32.8 g). Total weight and leaf weight both follow the same pattern, from highest to lowest yield: CFS, CF, IFS, IF, CT, and IT. For leaf weight and nitrogen dry weight percentage, the presence of fish water and sludge significantly ($p < 0.05$) enhanced the weight and nitrogen percentage for both compost and compost-free substrate; for height, this was only true for compost-free substrate. IT being last accords with what was observed for height, leaf biomass, and stem diameter; however, CFS and CF being first indicates that for this experiment, there is an additive effect of compost and fish effluents, with fish sludge performing better than fish water. This is to be expected, as fish sludge is much higher in all nutrients analysed (Table 1), with a 57% increase in nitrate content compared with fish water (44.3 mg/L vs. 69.6 mg/L). This is also reflected in the next two treatments that performed the best in terms of weight, IFS and IF; in these cases, fish sludge also performed better than fish water (average weight 38.125 g vs. 33 g). Second to last is treatment CT, where plants were grown in compost substrate and only watered with tap water; this indicates that fish water and fish sludge performed better than compost, in both total plant weight and total leaf weight.

The effects of fish effluents applied to soil for basil cultivation have only been investigated in two studies. Omeir, M.K. et al. [43] found that irrigation with fish farm effluent significantly increased the growth rates and nutrient content of the plants, compared to river water irrigation. Similarly, Valkovszki, N.J. et al. [44] reported that irrigation with fish farm effluents significantly increased the fresh and dry weight of shoots and roots, leaf number, and stem height in basil, indicating positive effects on plant growth.

The effect of fish effluent irrigation has also been investigated on another herb, oregano. Kimera, F. et al. [45] investigated the growth and essential oil content of *Origanum syriacum*. The results demonstrated that fish effluent irrigation significantly improved plant growth, with the treatment reaching maximum plant height and highest fresh and dry herbage yield. Although not performed in greenhouse conditions, these results are consistent with the results of this study, where the plants not fertilised with fish effluents (CT treatment) grew significantly less than the ones that were fertilised with fish effluents. No study so far, however, has compared the performance of compost and fish effluents for growing basil. Regarding the concentration of nitrogen in the leaves, fish sludge outperformed fish water in both substrate types, following the trend found for the leaf biomass, and reflected in the total plant fresh weight.

These results overall, and specifically the total fresh weight, are consistent with the literature on the positive effects of combining compost with fertiliser use. A study demonstrated that the combination of biochar compost and inorganic nitrogen fertiliser improved nitrogen uptake in rice plants [46]. Similarly, it was found that combining compost with organic fertiliser application in organic systems increased nutrient availability to plants in the early stages [47]. Furthermore, the combination of compost and nitrogen fertiliser improved yields, yield components, and nitrogen uptake in wheat cultivars [48]. These findings collectively support the notion that the combination of compost and fertiliser can enhance nitrogen uptake and concentration in plants. Moreover, a study revealed that the highest values for yield and its components were obtained by humic acid under the highest level of nitrogen fertiliser and compost treatment [49]. This suggests that the combination of compost and nitrogen fertiliser not only enhances nitrogen concentration, but also positively influences overall plant yield and components.

Regarding the microbiome communities found in the substrates, the final species count was higher at the end of the experiment than at the beginning in all treatments, including the ones where only tap water was supplied. This is probably due to the John

Innes seed substrate no. 1, where the seedlings were grown prior to being transferred to the pots. In fact, the measured initial diversity came from the manufactured substrates, which did not take into account the possible inoculation of further species derived from the tap water, air, or the seedling substrate where the basil seeds germinated. It is therefore likely that the remaining substrate surrounding the roots of the seedlings as they were being transferred into the experimental pots contributed to the overall microbial diversity, despite most of the substrate having been washed off. On the other hand, the soil microbiome has been extensively shown to mutate as plant growth progresses, its biodiversity having significant effects on the success and function of plant-associated microbiomes; thus, the increase in species diversity could have also occurred naturally alongside plant growth and development [50]. Compost has been documented as having rich microbial communities. Rastogi, M. et al. [51], for example, highlighted the role of microbes in solid waste composting, emphasising their vital contribution to the decomposition of organic matter and the production of soil-enriching compost. These microbes can, via compost, be introduced into the potting substrate. Green, S.J. et al. [52] investigated the succession of bacterial communities during early plant development and the transition from seed to root, emphasising the introduction of high numbers of microbial cells into soils or potting substrates through compost amendments. Michel, F.C. et al. [53] demonstrated the inoculation of compost-amended potting mixes with biocontrol agents and other microbes to induce systemic disease resistance in plants, indicating the potential for compost-derived microbes to influence soil and plant health. Furthermore, Zhao, J. et al. [54] screened and applied microbial inoculants for sewage sludge composting, highlighting the importance of understanding and utilising specific microbial communities in the composting process. Overall, the literature provides substantial evidence supporting the potential for microbes from compost to inoculate potting soil, influencing soil properties, plant growth, and overall soil health.

For treatments IF, IFS and CFS, the species count was higher halfway through the experiment than at the end, the difference being the lowest (5 more species) for IF, higher for IFS (35 more species), and highest for CFS (255 more species). This could be due to some initial measured increase in diversity derived from the seedling substrate, which then stabilised and decreased. The initial compost substrate was higher in diversity than the initial compost-free substrate (139 more species, or a 20.47% increase), most likely because of the enhancing effect of composted green material. This initial higher diversity in the compost substrate likely resulted in a higher microbial diversity in the treatments with compost substrate (CT, CF, CFS) than in the treatments with compost-free substrate (IT, IF, IFS); indeed, as the nutrients already present in the substrate combined with the nutrients provided by the fish effluents, in the same way, the microbial diversity supplied by the compost was enhanced by the supplementation of fish effluents. It has been found that potting soil mixed with compost can indeed lead to higher microbial diversity. The authors of [55] highlighted that composting leads to changes in bacterial and fungal communities, indicating an increase in microbial diversity. Neher, D.A. et al. [56] reported that compost can effectively replace peat-based potting media without negatively affecting plant growth, indicating the potential for compost to enhance microbial diversity in potting mixes. These findings are consistent with the study by [57], which shows that compost-based growing media can improve soil microbial activity. The introduction of compost into potting mixes was shown to enhance microbial diversity, as evidenced by the changes in bacterial and fungal communities, thus supporting the findings of this study.

Massilia, a major group of rhizosphere- and root-colonising bacteria [58], was present in the six most prevalent genera across all treatments. *Micromonospora*, a group of aerobic, mycelium-forming bacteria [59], was only present within the six most prevalent genera in the compost-free substrate treatments (IT, IF, and IFS), and *Lysobacter* and *Devosia* only in the treatments with compost substrate (CT, CF, and CFS). *Lysobacter* has been observed on the surface of fish tanks of aquaponic systems [60], as well as in RAS biofilter samples of aquaponic systems [61]; this genus has been identified as a plant growth-promoting bacterium (PGPB) that protects plants from disease through the production of antibiotics [62]. *Devosia* contains some potential nitrogen-fixing bacteria, and the genus has been found to grow on lettuce (*Lactuca sativa*) roots in aquaponic systems [60], as well as in all biofilters and sumps of an aquaponic system [61]. Taking only the most prevalent six genera into account per treatment, specific genera were only observed in treatments with compost substrate, and others only in treatments with compost-free substrate. This distinction is likely to have been caused by the initial substrate composition and the microbial communities associated with each. From the genera found, there does not seem to be a pattern associated with any specific treatment. In fact, in both bacterial and archaeal, and fungal analyses, most genera were found across multiple treatments. The impacts of fish effluents on the soil microbiome have not been extensively studied, especially not in greenhouse conditions. Some studies have however found that irrigation with fish effluents changes the structure of soil microbial communities. Chen, L. et al. [63] showed that aquaculture wastewater irrigation can change soil microbial functional diversity and community structure in arid regions, highlighting the potential impact of fish water irrigation on soil microbiota. Guan, W. et al. [64] found that the soil bacterial communities in rice fields irrigated with aquaculture wastewater can be impacted by the intestinal bacteria of the fish. Irrigation with fish-processing effluents has also been found to increase the nitrification rate and abundance of ammonia-oxidising archaea and bacteria in arid soils, but not to affect bacterial amoA genes [65]. Fish-processing effluent discharges in arid soils in Patagonia have also been shown to influence physicochemical properties and prokaryotic community structure, potentially benefitting native salt-tolerant plant irrigation [66]. However, Sun, Y. et al. [67] found that irrigation using activated brackish water improved soil fertility, which may have conferred dominant microbial populations with a strong ability to resist external interference to the soil environment, resulting in no significant impact on soil bacterial diversity. Given the limited knowledge of each particular microbial genus, it is difficult to assess the changes in the structure of the community, and how significant they were.

5. Conclusions

This experiment indicated the great potential of using fish water and fish sludge as fertiliser for greenhouse-based herb production. Fish effluents were able to fully substitute for compost, and to perform better than compost in some parameters, such as plant and leaf weight, thus showing that compost could be fully or partially substituted with fish effluents. The study therefore contributes valuable insights into the potential of fish effluents as a viable alternative to traditional compost in greenhouse agriculture. Whilst not without the need for further investigation, the results suggest a more sustainable approach to nutrient recycling and waste reduction in agricultural practices. The increase in microbe species count in all treatments, including IT where only tap water was used, was likely caused by the inoculation of microbiome species coming from the substrate where the seedlings were grown; inoculation could additionally have come from the tap water or ambient air, as the experiment was not performed in a sterile environment. The higher species diversity associated with the compost substrate treatments likely influenced the final diversity, given that a higher diversity was observed in the final samples of the compost substrate treatments than in the final samples of the compost-free substrate treatments.

In conclusion, this experiment aligns with existing literature by demonstrating that the presence of compost in growing media enhances microbial diversity. Regarding the different genera observed across treatments, for both bacteria and archaea, and for fungi, most genera were found across multiple treatments, without a clear genus pattern influenced by treatment. Given the higher microbial diversity observed following the application of the fish effluents, this experiment showed the potential for using fish effluents to create a 'living soil', which is one of the prerequisites for organic certification in Regulation (EU) 2018/848 (Annex II, rule 1.1) [1]. This study underscores the need for additional research to substantiate the implications of compost application, particularly in conjunction with fish effluent fertilisation, on soil microbial genera. Whilst the current findings suggest a positive influence on microbial diversity, a more comprehensive understanding of the nuanced dynamics necessitates further exploration of the intricate interactions between compost, fish effluent fertilisation, and microbial communities, elucidating the specific taxonomic composition and functional roles of microbial genera in response to these soil management practices. The conclusions of this study are based on a single growing cycle, which may not fully account for the long-term effects of repeated fish effluent application. Because organic fertilisers such as fish effluents can gradually shape soil fertility and microbial communities over multiple seasons, further research should include extended or multi-year studies. Such an approach would provide a more comprehensive evaluation of how fish effluents influence crop performance, soil quality, and ecological sustainability over time, thereby offering stronger evidence for their potential inclusion in organic production systems. Future research should also focus on understanding the best possible combination of fertilisation frequency, amount of fertiliser use, and type (filtered vs. unfiltered fish effluents), and investigating how much compost the effluents can truly substitute without compromising crop yield.

Author Contributions: Conceptualization, L.F.; methodology, L.F.; formal analysis, L.F. and M.P.P.; investigation, L.F. and B.K.; data curation, L.F.; writing—original draft preparation, L.F.; writing—review and editing, B.K., M.P.P. and S.M; supervision, B.K., M.P.P. and S.M.; project administration, L.F.; funding acquisition, B.K. All authors have read and agreed to the published version of the manuscript.

Funding: This study was funded by a University of Greenwich Vice-Chancellor's PhD Scholarship awarded to Lorenzo Fruscella.

Institutional Review Board Statement: Not applicable.

Informed Consent Statement: Not applicable.

Data Availability Statement: The raw data supporting the conclusions of this article will be made available by the authors on request.

Acknowledgments: The authors would like to thank Biome Makers Inc. for providing the analyses of the soil microbiome free of charge as part of the 'Fields4Ever' initiative, for the continuous support throughout the study, and for the data analyses provided.

Conflicts of Interest: The authors declare no conflicts of interest.

Appendix A

The following are the methods employed by NRM Labs to analyse the substrate and fish effluent samples.

Substrate—ammonium nitrate extractable calcium and sodium

The samples were air-dried at a temperature not exceeding 30 °C and sieved through a 2 mm mesh. The calcium and sodium were extracted by shaking the samples with M ammonium nitrate at 20 °C for 30 min. After filtration, the calcium concentration in the extract was measured by Atomic Absorption Spectrophotometry.

Substrate—available sulphate

The samples were air-dried at a temperature not exceeding 30 °C and sieved through a 2 mm mesh. The available sulphate was extracted under controlled conditions using a phosphate buffer (ratio 1:2). The filtered extract was then analysed by Inductively Coupled Plasma Emission Spectroscopy.

Substrate—DTPA-extractable manganese, iron, copper and zinc

The samples were air-dried at a temperature not exceeding 30 °C and sieved through a 2 mm mesh. Zinc, manganese, iron, and copper were extracted at 20 °C using a DTPA solution in a 1:2 ratio. In principle, the DTPA extraction allowed the metal in the samples to reach equilibrium with the chelating agent. A pH of 7.3 enabled DTPA to extract iron as well as other metals.

Substrate—hot water-soluble boron

The samples were air-dried at a temperature not exceeding 30 °C and sieved through a 2 mm mesh. Boron availability was assessed by hot water extraction. The boron concentration in the extract was measured using ICP-OES (Inductively Coupled Plasma Optical Emission Spectroscopy).

Substrate—organic matter content by loss on ignition at 430 °C

The samples were air-dried at a temperature not exceeding 30 °C and sieved through a 2 mm mesh. The organic matter was destroyed by dry combustion at 430 °C, and the loss in weight of the samples was reported as the organic matter content (in %) on a dry matter basis.

Substrate—mineral nitrogen (available N)

The samples were chopped and mixed to achieve a homogeneous sample. A portion was shaken with 2 M KCl to extract the mineral-N fractions, and a dry matter determination was performed. If mineralisable N was required, a second portion was incubated anaerobically at 40 °C in water for one week to mineralise the nitrogen, followed by extraction of the mineralised fraction with KCl. Once in solution, nitrate-N, nitrite-N, and ammonium-N were measured colourimetrically. Nitrate-N and nitrite-N were determined based on the formation of a diazo compound between nitrite and sulphanilamide, which was then coupled with N-1-naphthylethylenediamine dihydrochloride to form a red azo dye. The colour was measured at 540 nm. In channel one, nitrate was fully reduced to nitrite by cadmium metal in an open tubular cadmium reactor (OTCR), so total oxidised nitrogen (TON) was measured as the sum of nitrite plus reduced nitrate. In channel two, only nitrite was measured. Nitrate-N was calculated by subtracting the nitrite value from the TON. For ammonium-N, in channel three, ammonium reacted with alkaline hypochlorite and phenol to produce indophenol blue. Sodium nitroprusside served as a catalyst in the formation of indophenol blue, which was measured at 640 nm. Precipitation of calcium and magnesium hydroxides was prevented by adding a combined potassium sodium tartrate/sodium citrate complexing reagent.

Substrate—pH and lime requirement

The samples were air-dried at a temperature not exceeding 30 °C and sieved through a 2 mm mesh. pH was defined as the pH, measured potentiometrically, of a suspension created by stirring the samples with water in a 1:2.5 ratio. As temperature influenced pH measurement, the measurement was performed in a temperature-controlled setting.

Substrate—Olsen's extractable phosphorus

The samples were air-dried at a temperature not exceeding 30 °C and sieved through a 2 mm mesh. The available phosphorus was extracted at 20 °C by shaking the samples with 0.5 M sodium bicarbonate for 30 min. The phosphorus concentration was then determined by flow injection analysis/colourimetry, whereby acid ammonium molybdate formed the phosphomolybdate ion. This ion, when reduced with ascorbic acid, created a blue complex, measured spectrophotometrically at 880 nm. Calibration was carried out with commercial phosphate standards traceable to SI units.

Substrate—ammonium nitrate-extractable potassium and magnesium

Potassium and magnesium were extracted at 20 °C by shaking the samples with 1 M ammonium nitrate for 30 min. After filtration, the concentrations of potassium and magnesium in the extract were determined by Atomic Absorption Spectrometry, calibrated with commercial standards traceable to SI units.

Fish effluent—nitrate nitrogen

Nitrate-N was measured colourimetrically based on the formation of a diazo compound between nitrite and sulphanilamide, which was subsequently coupled with N-1-naphthylethylenediamine dihydrochloride to form a red azo dye, measured at 540 nm. In channel one, nitrate was fully reduced to nitrite by cadmium metal in an open tubular cadmium reactor (OTCR). Thus, total oxidised nitrogen (TON) was measured as the combination of nitrite plus reduced nitrate. In channel two, nitrite alone was measured. Nitrate-N was calculated by subtracting nitrite from the TON.

Fish effluent—dissolved elements

The samples were filtered to remove particulate matter. The elements in the filtrate were then measured either by Inductively Coupled Plasma Mass Spectroscopy (ICP-MS) or by Inductively Coupled Plasma Emission Spectroscopy (ICP-OES), depending on the specific element of interest and the required Limit of Detection (LOD). ICP-MS enabled lower LODs.

Fish effluent—electrical conductivity

The specific conductivity of the solution was determined using an EC meter, and the measurement was standardised to 25 °C.

Fish effluent—pH

The pH of the solution was measured potentiometrically. As temperature influenced the result, the measurement was performed in a temperature-controlled environment.

Fish effluent—total dissolved solids

After filtration through glass fibre paper to remove suspended solids, a known volume was dried at 180 °C and the residue weighed in order to calculate the total dissolved solids.

References

1. Regulation (EU) 2018/848 of the European Parliament and of the Council of 30 May 2018 on organic production and labelling of organic products and repealing Council Regulation (EC) No 834/2007. Available online: https://eur-lex.europa.eu/eli/reg/2018/848/oj/eng (accessed on 31 March 2024).
2. Commission Implementing Regulation (EU) 2023/121 of 17 January 2023 amending and correcting Implementing Regulation (EU) 2021/1165 authorising certain products and substances for use in organic production and establishing their lists. Available online: https://eur-lex.europa.eu/eli/reg_impl/2023/121/oj/eng (accessed on 31 March 2024).
3. Rakocy, J.E.; Bailey, D.S.; Shultz, R.C.; Thoman, E.S. Update on tilapia and vegetable production in the UVI aquaponic system. In *New Dimensions on Farmed Tilapia, Proceedings of the Sixth International Symposium on Tilapia in Aquaculture, Manila, Philippines, 12–16 September 2004*; Bolivar, R., Mair, G., Fitzsimmons, K., Eds.; Creative Unlimited: Clovelly, NSW, Australia, 2004; pp. 298–312.

4. Graber, A.; Junge, R. Aquaponic systems: Nutrient recycling from fish wastewater by vegetable production. *Desalination* **2009**, *246*, 147–156. [CrossRef]
5. Tyson, R.V.; Treadwell, D.D.; Simonne, E.H. Opportunities and challenges to sustainability in aquaponic systems. *HortTechnology* **2011**, *21*, 6–13. [CrossRef]
6. Naylor, S.J.; Moccia, R.D.; Durant, G.M. The chemical composition of settleable solid fish waste (manure) from commercial rainbow trout farms in Ontario, Canada. *N. Am. J. Aquac.* **1999**, *61*, 21–26. [CrossRef]
7. Castro, R.S.; Borges Azevedo, C.M.S.; Bezerra-Neto, F. Increasing cherry tomato yield using fish effluent as irrigation water in Northeast Brazil. *Sci. Hortic.* **2006**, *110*, 44–50. [CrossRef]
8. Gravel, V.; Dorais, M.; Dey, D.; Vandenberg, G. Fish effluents promote root growth and suppress fungal diseases in tomato transplants. *Can. J. Plant Sci.* **2015**, *95*, 427–436. [CrossRef]
9. Mangmang, J.S.; Deaker, R.; Rogers, G. *Azospirillum brasilense* enhances recycling of fish effluent to support growth of tomato seedlings. *Horticulturae* **2015**, *1*, 14–26. [CrossRef]
10. Delaide, B.; Teerlinck, S.; Decombel, A.; Bleyaert, P. Effect of wastewater from a pikeperch (*Sander lucioperca* L.) recirculated aquaculture system on hydroponic tomato production and quality. *Agric. Water Manag.* **2019**, *226*, 105814. [CrossRef]
11. Pattillo, A.D.; Foshee, W.G.; Blythe, E.K.; Pickens, J.; Wells, D.; Monday, T.A.; Hanson, T.R. Performance of aquaculture effluent for tomato production in outdoor raised beds. *HortTechnology* **2020**, *30*, 624–631. [CrossRef]
12. Pickens, J.M.; Danaher, J.J.; Sibley, J.L.; Chappell, J.A.; Hanson, T.R. Integrating greenhouse cherry tomato production with biofloc tilapia production. *Horticulturae* **2020**, *6*, 44. [CrossRef]
13. Diatta, A.A.; Manga, A.G.B.; Bassène, C.; Mbow, C.; Battaglia, M.; Sambou, M.; Babur, E.; Uslu, Ö.S. Sustainable production of tomato using fish effluents improved plant growth, yield components, and yield in northern Senegal. *Agronomy* **2023**, *13*, 2696. [CrossRef]
14. Mechouma, A.; Mezerdi, F. Impact of the use of water from fish farming in the irrigation of pepper crops (*Capsicum annuum* L.) in the greenhouse in the Biskra region. *Int. J. Environ. Stud.* **2024**, *81*, 734–750. [CrossRef]
15. Lenz, G.L.; Loss, A.; Lourenzi, C.R.; Lopes, D.L.A.; Siebeneichler, L.M.; Brunetto, G. Common chicory production in aquaponics and in soil fertilized with aquaponics sludge. *Sci. Hortic.* **2021**, *281*, 109946. [CrossRef]
16. Mangmang, J.S.; Deaker, R.S.; Rogers, G. Response of lettuce seedlings fertilized with fish effluent to *Azospirillum brasilense* inoculation. *Biol. Agric. Hortic.* **2015**, *31*, 61–71. [CrossRef]
17. Goddek, S.; Schmautz, Z.; Scott, B.; Delaide, B.; Keesman, K.J.; Wuertz, S.; Junge, R. The effect of anaerobic and aerobic fish sludge supernatant on hydroponic lettuce. *Agronomy* **2016**, *6*, 37. [CrossRef]
18. Delaide, B.; Panana, E.; Teerlinck, S.; Bleyaert, P. Suitability of supernatant of aerobic and anaerobic pikeperch (*Sander lucioperca* L.) sludge treatments as a water source for hydroponic production of lettuce (*Lactuca sativa* L. var. capitata). *Aquac. Int.* **2021**, *29*, 1721–1735. [CrossRef]
19. Lenz, G.L.; Loss, A.; Lourenzi, C.R.; Lopes, D.L.A.; Siebeneichler, L.M.; Brunetto, G. Lettuce growth in aquaponic system and in soil fertilized with fish sludge. *Aquac. Res.* **2021**, *52*, 5008–5021. [CrossRef]
20. Mangmang, J.S.; Deaker, R.S.; Rogers, G. Response of cucumber seedlings fertilized with fish effluent to *Azospirillum brasilense*. *Int. J. Veg. Sci.* **2016**, *22*, 129–140. [CrossRef]
21. Abdelraouf, R.E. Reuse of fish farm drainage water in irrigation. In *Unconventional Water Resources and Agriculture in Egypt*; Negm, A., Ed.; Springer: Cham, Switzerland, 2017; pp. 393–410.
22. Chen, S.; Coffin, D.E.; Malone, R.F. Sludge production and management for recirculating aquacultural systems. *J. World Aquac. Soc.* **1997**, *28*, 303–315. [CrossRef]
23. Montanhini Neto, R.; Ostrensky, A. Nutrient load estimation in the waste of Nile tilapia *Oreochromis niloticus* (L.) reared in cages in tropical climate conditions. *Aquac. Res.* **2013**, *46*, 1309–1322. [CrossRef]
24. Brod, E.; Oppen, J.; Kristoffersen, A.Ø.; Haraldsen, T.K.; Krogstad, T. Drying or anaerobic digestion of fish sludge: Nitrogen fertilisation effects and logistics. *Ambio* **2017**, *46*, 852–864. [CrossRef]
25. Rafiee, G.; Saad, C.R. Nutrient cycle and sludge production during different stages of red tilapia (Oreochromis sp.) growth in a recirculating aquaculture system. *Aquaculture* **2005**, *244*, 109–118. [CrossRef]
26. Cripps, S.J.; Bergheim, A. Solids management and removal for intensive land-based aquaculture production systems. *Aquac. Eng.* **2000**, *22*, 33–56. [CrossRef]
27. Monsees, H.; Keitel, J.; Paul, M.; Kloas, W.; Wuertz, S. Potential of aquacultural sludge treatment for aquaponics: Evaluation of nutrient mobilization under aerobic and anaerobic conditions. *Aquac. Environ. Interact.* **2017**, *9*, 9–18. [CrossRef]
28. Wotton, R.; Malmqvist, B. Feces in aquatic ecosystems. *BioScience* **2001**, *51*, 537–544. [CrossRef]
29. Kledal, P.R.; König, B.; Matulić, D. Aquaponics: The ugly duckling in organic regulation. In *Aquaponics Food Production Systems*; Goddek, S., Joyce, A., Kotzen, B., Burnell, G., Eds.; SpringerOpen: Cham, Switzerland, 2019; pp. 487–500. [CrossRef]
30. Rahimi, S.; Modin, O.; Mijakovic, I. Technologies for biological removal and recovery of nitrogen from wastewater. *Biotechnol. Adv.* **2020**, *43*, 107570. [CrossRef]

31. Vaccari, D.A. Phosphorus: A looming crisis. *Sci. Am.* **2009**, *300*, 54–59. [CrossRef]
32. Scholz, R.W.; Ulrich, A.E.; Eilittä, M.; Roy, A. Sustainable use of phosphorus: A finite resource. *Sci. Total Environ.* **2013**, *461–462*, 799–803. [CrossRef]
33. Rakocy, J.; Shultz, R.C.; Bailey, D.S.; Thoman, E.S. Aquaponic production of tilapia and basil: Comparing a batch and staggered cropping system. *Acta Hortic.* **2004**, *648*, 63–69. [CrossRef]
34. Xie, K.; Rosentrater, K. Life cycle assessment (LCA) and techno-economic analysis (TEA) of tilapia-basil aquaponics. In Proceedings of the ASABE Annual International Meeting, New Orleans, LA, USA, 26–29 July 2015. Available online: https://dr.lib.iastate.edu/server/api/core/bitstreams/566003c3-9fe9-4735-a8c8-b4dfa8a4c32c/content (accessed on 31 March 2024).
35. Ferrarezi, R.S.; Bailey, D.S. Basil performance evaluation in aquaponics. *HortTechnology* **2019**, *29*, 85–93. [CrossRef]
36. Knaus, U.; Pribbernow, M.; Xu, L.; Appelbaum, S.; Palm, H.W. Basil (*Ocimum basilicum*) cultivation in decoupled aquaponics with three hydro-components (grow pipes, raft, gravel) and African catfish (*Clarias gariepinus*) production in Northern Germany. *Sustainability* **2020**, *12*, 8745. [CrossRef]
37. Yang, T.; Kim, H.-J. Characterizing nutrient composition and concentration in tomato-, basil-, and lettuce-based aquaponic and hydroponic systems. *Water* **2020**, *12*, 1259. [CrossRef]
38. Pasch, J.; Ratajczak, B.; Appelbaum, S.; Palm, H.W.; Knaus, U. Growth of basil (*Ocimum basilicum*) in DRF, raft, and grow pipes with effluents of African catfish (*Clarias gariepinus*) in decoupled aquaponics. *Agric. Eng.* **2021**, *3*, 92–109. [CrossRef]
39. JohnnySeeds. Hydroponic & Container Basil Variety Comparison Charts. Available online: https://www.johnnyseeds.com/growers-library/herbs/basil/hydroponic-container-basil-varieties-comparison-chart.html (accessed on 31 March 2024).
40. Royal Horticultural Society. Onions. Available online: https://www.rhs.org.uk/advice/grow-your-own/vegetables/onions (accessed on 31 March 2024).
41. RStudio Team. RStudio: Integrated Development for R. Available online: https://posit.co/download/rstudio-desktop/ (accessed on 31 March 2024).
42. Fruscella, L.; Kotzen, B.; Paradelo, M.; Milliken, S. Investigating the effects of fish effluents as organic fertilisers on onion (Allium cepa) yield, soil nutrients, and soil microbiome. *Sci. Hortic.* **2023**, *321*, 112297. [CrossRef]
43. Omeir, M.K.; Jafari, A.; Shirmardi, M.; Roosta, H. Effects of irrigation with fish farm effluent on nutrient content of basil and purslane. *Proc. Natl. Acad. Sci. India B* **2020**, *90*, 825–831. [CrossRef]
44. Valkovszki, N.J.; Jancsó, M.; Székely, Á.; Szalóki, T.; Kolozsvári, I.; Kun, Á. Influence of agricultural effluent irrigation on common purslane (*Portulaca oleracea* l.) and garden basil (*Ocimum basilicum* L.): Preliminary results. *J. Agric. Environ. Sci.* **2022**, *9*, 71–81. [CrossRef]
45. Kimera, F.; Sewilam, H.; Fouad, W.M.; Suloma, A. Efficient utilization of aquaculture effluents to maximize plant growth, yield, and essential oils composition of Origanum majorana cultivation. *Ann. Agric. Sci.* **2021**, *66*, 1–7. [CrossRef]
46. Aboagye, D.A.; Adjadeh, W.T.; Nartey, E.K.; Asuming-Brempong, S. Co-application of biochar compost and inorganic nitrogen fertilizer affects the growth and nitrogen uptake by lowland rice in northern Ghana. *Nitrogen* **2022**, *3*, 414–425. [CrossRef]
47. Bi, G.; Li, T.; Gu, M.; Evans, W.B.; Williams, M. Effects of fertilizer source and rate on zinnia cut flower production in a high tunnel. *Horticulturae* **2021**, *7*, 333. [CrossRef]
48. Al-Dulaimi, O.I.M.; Al-Rawi, A.R.M.; Al-Qaisi, E.K.K.; El-Moursy, R.S.A. Response of some wheat cultivars to organic, mineral and foliar fertilization. *J. Plant Prod.* **2015**, *6*, 1755–1770. [CrossRef]
49. Antoun, L.W.; Zakaria, S.M.; Rafla, H.H. Influence of compost, N-mineral and humic acid on yield and chemical composition of wheat plants. *J. Soil Sci. Agric. Eng.* **2010**, *1*, 1131–1143. [CrossRef]
50. Saleem, M.; Hu, J.; Jousset, A. More than the sum of its parts: Microbiome biodiversity as a driver of plant growth and soil health. *Annu. Rev. Ecol. Evol. Syst.* **2019**, *50*, 145–168. [CrossRef]
51. Rastogi, M.; Nandal, M.; Khosla, B. Microbes as vital additives for solid waste composting. *Heliyon* **2020**, *6*, e03343. [CrossRef] [PubMed]
52. Green, S.J.; Inbar, E.; Michel, F.C.; Hadar, Y.; Minz, D. Succession of bacterial communities during early plant development: Transition from seed to root and effect of compost amendment. *Appl. Environ. Microbiol.* **2006**, *72*, 3975–3983. [CrossRef] [PubMed]
53. Michel, F.C.; Hoitink, H.A.J.; Hadar, Y.; Minz, D. *Microbial Communities Active in Soil Induced Systemic Plant Disease Resistance*; United States Department of Agriculture: Washington, DC, USA, 2005.
54. Zhao, J.; Wang, X.W.; Fan, H.; Fan, C.P.; Shao, S.G.; Xie, J.B. Screening and application of microbial inoculants for sludge composting in expressway service area of northwest China. *E3S Web Conf.* **2021**, *248*, 01037. [CrossRef]
55. Partanen, P.; Hultman, J.; Paulín, L.; Auvinen, P.; Romantschuk, M. Bacterial diversity at different stages of the composting process. *BMC Microbiol.* **2010**, *10*, 94. [CrossRef]
56. Neher, D.A.; Weicht, T.R.; Bates, S.T.; Leff, J.W.; Fierer, N. Changes in bacterial and fungal communities across compost recipes, preparation methods, and composting times. *PLoS ONE* **2013**, *8*, e79512. [CrossRef]

57. Lehman, R.M.; Cambardella, C.A.; Stott, D.E.; Acosta Martinez, V.; Manter, D.K.; Buyer, J.S.; Maul, J.E.; Smith, J.L.; Collins, H.P.; Halvorson, J.J.; et al. Understanding and enhancing soil biological health: The solution for reversing soil degradation. *Sustainability* **2015**, *7*, 988–1027. [CrossRef]
58. Ofek, M.; Hadar, Y.; Minz, D. Ecology of root colonizing Massilia (Oxalobacteraceae). *PLoS ONE* **2012**, *7*, e40117. [CrossRef]
59. Yan, S.; Zeng, M.; Wang, H.; Zhang, H. Micromonospora: A prolific source of bioactive secondary metabolites with therapeutic potential. *J. Med. Chem.* **2022**, *65*, 8735–8771. [CrossRef]
60. Schmautz, Z.; Walser, J.-C.; Espinal, C.A.; Gartmann, F.; Scott, B.; Pothier, J.F.; Frossard, E.; Junge, R.; Smits, T.H.M. Microbial diversity across compartments in an aquaponic system and its connection to the nitrogen cycle. *Sci. Total Environ.* **2022**, *852*, 158426. [CrossRef]
61. Eck, M.; Sare, A.R.; Massart, S.; Schmautz, Z.; Junge, R.; Smits, T.H.M.; Jijakli, M.H. Exploring bacterial communities in aquaponic systems. *Water* **2019**, *11*, 260. [CrossRef]
62. Reichenbach, H. The Lysobacter genus. In *The Prokaryotes*; Dworkin, M., Falkow, S., Rosenberg, E., Schleifer, K.H., Stackebrandt, E., Eds.; Springer: New York, NY, USA, 2006; pp. 939–957.
63. Chen, L.; Feng, Q.; Li, C.; Wei, Y.; Zhao, Y.; Feng, Y.; Zheng, H.; Li, F.; Li, H. Impacts of aquaculture wastewater irrigation on soil microbial functional diversity and community structure in arid regions. *Sci. Rep.* **2017**, *7*, 11193. [CrossRef]
64. Guan, W.; Li, K.; Li, K. Bacterial communities in co-cultured fish intestines and rice field soil irrigated with aquaculture wastewater. *AMB Express* **2022**, *12*, 132. [CrossRef] [PubMed]
65. Marcos, M.S.; González, M.C.; Vallejos, M.B.; Barrionuevo, C.G.; Olivera, N.L. Impact of irrigation with fish-processing effluents on nitrification and ammonia-oxidizer abundances in Patagonian arid soils. *Arch. Microbiol.* **2021**, *203*, 3945–3953. [CrossRef] [PubMed]
66. Vallejos, M.B.; Marcos, M.S.; Barrionuevo, C.; Olivera, N.L. Fish-processing effluent discharges influenced physicochemical properties and prokaryotic community structure in arid soils from Patagonia. *Sci. Total Environ.* **2020**, *714*, 136882. [CrossRef]
67. Sun, Y.; Wang, C.; Mi, W.; Qu, Z.; Mu, W.; Wang, J.; Zhang, J.; Wang, Q. Effects of irrigation using activated brackish water on the bacterial community structure of rhizosphere soil. *J. Soil Sci. Plant Nutr.* **2022**, *22*, 4008–4023. [CrossRef]

applied sciences

MDPI

Article

^{15}N-Nitrogen Use Efficiency, Productivity, and Quality of Durum Wheat Integrating Nitrogen Management and an Indigenous Bacterial Inoculant in a Single Growing Season

Marisol Ayala-Zepeda [1], Fannie Isela Parra-Cota [2], Cristina Chinchilla-Soto [3], Eulogio De La Cruz-Torres [4], María Itria Ibba [5], María Isabel Estrada-Alvarado [6] and Sergio de los Santos-Villalobos [1,*]

1 Departamento de Ciencias Agronómicas y Veterinarias, Instituto Tecnológico de Sonora ITSON, Obregón 85000, Mexico; marisol.ayala104893@potros.itson.edu.mx
2 Campo Experimental Norman E. Borlaug, Instituto Nacional de Investigaciones Forestales, Agrícolas y Pecuarias INIFAP, Obregón 85000, Mexico; parra.fannie@inifap.gob.mx
3 Centro de Investigación en Contaminación Ambiental, Universidad de Costa Rica UCR, San José 11501-2060, Costa Rica; cristina.chinchilla@ucr.ac.cr
4 Departamento de Biología, Instituto Nacional de Investigaciones Nucleares ININ, Ocoyoacac 52750, Mexico; eulogio.delacruz@inin.gob.mx
5 Centro Internacional de Mejoramiento de Maíz y de Trigo CIMMYT, Texcoco 56237, Mexico; m.ibba@cgiar.org
6 Departamento de Biotecnología y Ciencias Alimentarias, Instituto Tecnológico de Sonora ITSON, Obregón 85000, Mexico; mestrada@itson.edu.mx
* Correspondence: sergio.delossantos@itson.edu.mx

Abstract: In the Yaqui Valley, Mexico, the current and estimated annual growth rate of durum wheat (*Triticum turgidum* L. subsp. *durum*) yield is insufficient to satisfy the food demand that the world will be facing by 2050. Furthermore, besides the high doses of fertilizers applied to wheat in this region, nitrogen-use efficiency (NUE) remains low (<34.4%). A sustainable strategy to reduce the use of fertilizers and to increase crop yield and quality is the use of native plant growth-promoting bacteria as microbial inoculants. This study was performed under field conditions during one agricultural season in the Yaqui Valley, Mexico. It aimed to quantify the impact of the inoculation of a native bacterial consortium (BC) composed of *Bacillus cabrialesii* subsp. *cabrialesii* TE3^{T}, *Priestia megaterium* TRQ8, and *Bacillus paralicheniformis* TRQ65 on grain yield, grain quality, and NUE (measured through ^{15}N-isotopic techniques) at different stages of development of durum wheat variety CIRNO C2008 under three doses of urea (0, 120, and 240 kg N ha^{-1}) fractionated at 30%, 60%, and 10%. Results showed that yield, quality, and NUE were highly affected by the N doses, while the inoculation of the BC had a lower impact on these parameters. Nevertheless, the inoculation of the BC on wheat had positive effects at the early stages of growth, on plant height (+1.6 cm), root depth (+11.9 cm), and spikes per square meter (+25 spikes m^{-2}). Moreover, the addition of the BC improved N acquisition by the plants, at different crop growth stages, compared to uninoculated treatments. Finally, our results indicated that reducing the N dose from 240 kg of N ha^{-1} to 120 kg of N ha^{-1} improved the NUE (27.5% vs. 44.3%, respectively) of the crop. Hence, results of this preliminary study showed that the incorporation of bacterial inoculants into the wheat crop requires a simultaneous adequate N management, in order to obtain the desired positive effect on wheat productivity.

Keywords: bacterial consortium; wheat yield; sustainable agriculture; *Bacillus*; *Priestia*

1. Introduction

Wheat and its derivates, among all crops, are the main source of calories and nutrients to the human diet, grown in 70% of the world's cultivated area, across a wide range of edaphoclimatic conditions, which makes it the most significant grain crop worldwide [1,2]. The Yaqui Valley (birthplace of the Green Revolution) and the Mayo Valley are the main farming regions of the State of Sonora, Mexico. In 2023, a total area of 253,414 hectares of wheat was planted in these valleys, which has an average production of 1,893,260 tons, representing 56% of national wheat production [3]. However, nitrogen-use efficiency (NUE) of the wheat crop in this region has been reported in a range from 25.3 to 34.4% [4,5]. The information above suggests that most of the N applied is lost in the environment through runoff, leaching, deposition, volatilization, and/or denitrification. These cause alterations in the ecosystems and biodiversity, and societal threats related to water, soil, and air quality, greenhouse gas imbalance, adverse health conditions, and economic losses [6].

Considering the demands of the growing population, and the fact that 30% of the costs of wheat production (2022–2023) in Mexico is used in fertilization [7], it is necessary to optimize the use of synthetic fertilizers applied to this crop, while increasing its yield, quality, and NUE [8]. Aiming to satisfy the 2050 estimated food demand, wheat yield needs to increase at an annual rate of 2.4%, but the global average rate of wheat yield increase was only 0.9% per year by 2013, while in Mexico, the annual rate for wheat increment was 1.1% [9]. Moreover, yield growth for the next 20 years in the Yaqui Valley is estimated to be 1.2% per year in the absence of sustainable technologies [10].

A sustainable and promising strategy to enhance plant growth, while increasing the dependency on synthetic fertilizers and pesticides is the use of plant growth-promoting bacteria (PGPB) [11]. These bioproducts have the potential to reduce the use of nitrogen, phosphorus, and potassium fertilizers by 50% and to increase crop yield by up to ~10 to 30% according to recent studies [12]. These bacteria can promote plant growth through direct mechanisms: (i) phytohormones modulation/production; (ii) 1-Aminocyclopropane-1-carboxylate deaminase (ACC deaminase) synthesis; (iii) nitrogen fixation; and (iv) phosphate solubilization, as well as through indirect mechanisms, such as (i) antibiotic production; (ii) lytic enzyme production; (iii) niche competition; (iv) endotoxin production; and (v) mycoparasitism, or mechanisms with dual effects, such as (i) phytohormones production; (ii) siderophores production; and (iii) quorum sensing [13].

To investigate the impact of inoculation and co-inoculation of rhizobacteria on durum wheat growth and productivity in alkaline calcareous soil, Chami et al. [14] established and compared four treatments during two agricultural cycles: an uninoculated control, inoculation on wheat with a *Paenibacillus polymyxa* SGH1 strain, inoculation with *P. polymyxa* SGK2, and their co-inoculation on wheat. Results showed that the co-inoculation of durum wheat with *P. polymyxa* strains had positive effects on wheat morphological traits (collar diameter, +16.9%; tillers plant^{-1}, +89.8%; weight of rhizospheric soil/dry weight of roots, +35.5%), yield (+41.1% in the first season and +16.6% in the second one) parameters, and quality traits, compared to uninoculated treatments, as a result of a synergistic activity [14].

The three native bacterial strains with plant growth-promoting traits used in the present experiment were isolated from the Yaqui Valley and have been studied by our research team under laboratory and field conditions. *Bacillus cabrialesii* subsp. *cabrialesii* TE3^{T} is an endophytic strain isolated from wheat's leaf tissue and reported as a novel bacterial species [15]; *Priestia megaterium* TRQ8 (previously affiliated and reported as *Bacillus megaterium*) and *Bacillus paralicheniformis* TRQ65 [16] were isolated from the rhizosphere of wheat in commercial wheat fields. Rojas Padilla et al. (2020) [17] reported that TRQ8 was able to produce siderophores (production index of 8.17). They also determined the solubilization index of insoluble phosphorus by TRQ8, TRQ65, and TE3^{T}, which resulted in ranges from 1.37 to 1.43. Finally, they quantified the strains' capacity to produce indoles, with amounts that ranged from 8.21 to 39.29 μg/mL.

Moreover, it was demonstrated that the co-inoculation of TRQ8 and TRQ65 in wheat increased their ability to promote growth in the early stages of development, without antagonistic effects among them. Thus, their co-inoculation showed significant ($p < 0.05$) increments on aerial length = 5.5%, root length = 10.5%, dry weight of the aerial part = 60.7%, root dry weight = 82.3%, and biovolume = 17.6% [17].

In addition, Rojas-Padilla et al. (2022) [18], conducted a greenhouse assay, where wheat seedlings were inoculated with alginate microbeads containing these strains, this treatment improved biometric parameters (vs. the non-inoculated treatment), such as root dry weight (13–77%), root length (8–19%), stem dry weight (31%), stem length (8–12%), and chlorophyll content (7%).

Field assays over four wheat crop cycles in the Yaqui Valley demonstrated that inoculating the consortium with the strains TRQ8, TRQ65, and TE3^{T}, and combined with different fertilization doses, increased grain yield by 0.5 to 2.0 ton ha^{-1} compared to non-inoculated treatments, while maintaining grain quality required by the industry [5,19]. In those studies, N fertilization fractionated into three equal parts (33.3%). However, current recommendations for wheat production in the Yaqui Valley are to apply N fertilization in splits: 30%, 60%, and 10%; 30%, 50%, and 20%, or 30, 70%, and 0% (in sowing, stem elongation, and booting), among others [20], and it is crucial to assess whether these N management recommendations are compatible with the use of native microorganisms as bacterial inoculants. Thus, this one-crop season study aimed to quantify the impact of inoculating a bacterial consortium (BC) (*Bacillus cabrialesii* subsp. *cabrialesii* TE3^{T}, *Priestia megaterium* TRQ8, and *Bacillus paralicheniformis* TRQ65) under three urea doses (0, 120, and 240 kg N ha^{-1}) fractionated at 30%, 60%, and 10%, on grain yield, grain quality, and NUE—through ^{15}N-isotopic techniques—at different stages of development of durum wheat (CIRNO C2008) in the Yaqui Valley, Mexico. It is crucial to assess the impact of inoculation on a specific N management strategy in the short term, highlighting the key findings for the studied traits. This is particularly relevant given that farmers are looking for beneficial effects within a single growing season. This study serves as a starting point for establishing broader, longer-term lines of research. The present findings can be used to adjust the experimental design in subsequent cycles.

2. Materials and Methods

2.1. Edapho-Climatic Characteristics of the Study Site and Agronomic Management Conditions

The study was conducted during the fall–winter crop season (2 December 2021 to 22 May 2022) at the Experimental Technology Transfer Center (CETT-910) of the Instituto Tecnológico de Sonora (ITSON) in the Yaqui Valley, Mexico (Latitude: 27°21′57.74″ N, Longitude: 109°54′55.91″ W). Climatic conditions during the crop cycle included 589 accumulated chill hours, an average temperature of 18.6 °C (mean maximum temperature of 29.1 °C, and mean minimum temperature of 9.5 °C, see Supplementary Table S1), total precipitation of 17.5 mm, and average relative humidity of 62.8% (REMAS, https://siafeson.com/remas/ accessed on 1 June 2022).

The physicochemical characteristics of the soil, macro, and micronutrients are shown in Table 1. These values were determined following the protocols outlined in the Official Mexican Standard NOM-021-SEMARNAT-2000. The nitrogen level reported in Table 1 was calculated from the soil sample analysis at UC-Davis (described in Section 2.5) using an Elemental Analyzer coupled with an Isotope Ratio Mass Spectrometer (EA-IRMS, Elementar Analysensysteme GmbH, Hanau, Germany). Total nitrogen was estimated considering the %N in the soil samples, converted into kg of N per hectare; accordingly considering the volume, bulk density, and the weight of the soil in 1 hectare. Finally, inorganic nitrogen (the available form of N) was calculated by multiplying each value by 0.035, as 3 to 4% of the total N in the Yaqui Valley soils is typically inorganic. The reported value represents an average of 15 samples.

Table 1. Physicochemical and elemental soil analysis.

Property	Value
Texture	Clay soil (sand 45.4%, silt 17.1%, clay 37.5%)
Water content at saturation	48%
Electrical conductivity	1.8 dS m^{-1}
pH	7.9
Organic matter content	1.00%
Calcium carbonates	0.10%
Hydraulic conductivity	1.0 cm h^{-1}
Bulk density (30 cm)	1.2 g cm^{-3}
Macronutrients	**Concentration (kg ha^{-1})**
Inorganic N	183
P	52.7
Mg	2509.2
Ca	22,032
K	2376
Na	3024
Cation exchange capacity	41.8 mEq 100 g^{-1}
Micronutrients	**Concentration (ppm)**
Fe	7.2
Mn	15.4
Zn	0.5
Cu	1.2
B	0.7

The land was prepared using fallow and primary tillage, which included chiseling and plowing. Durum wheat (*Triticum turgidum* L. subsp. *durum*), specifically the commercial cultivar CIRNO C2008, was sown on 2 December 2021, using a seeder (SUB-24) with three rows per furrow, at a seed density of 120 kg ha^{-1}, under field conditions, following the common wheat crop management of the Yaqui Valley on a vertisol soil—without crop rotation.

Disease, insect, and weed management were carried out as follows (all products were diluted in 40 L of water and applied using a backpack pesticide sprayer): for weed control (broadleaf herbs), 2 L ha^{-1} of Paraquat 30.1% (Gramoxone SL 2.0, Syngenta, Basel, Switzerland) were applied 5 days before planting; once the crop germinated (13 days after sowing), a mixture of 2,4-D dimethylamine salt 70.53% (Amina 6, AgroLucava, Celaya, Mexico) (1 L ha^{-1}) and Fluroxypyr: 1-methylheptyl ester 45.52% (Starane, Corteva, Seville, Spain) (0.5 L ha^{-1}) was applied. For the control of aphids, 80 days after sowing: Imidacloprid 18.8% + Lambda Cyhalotrin 13.6% (Corax SC, Lapisa, La Piedad, Mexico) (200 mL ha^{-1}), and to control leaf rust: Propiconazole 25.5% (Sanazole 250 EC, Velsimex, Mexico City, Mexico) (600 mL ha^{-1}).

Mono-ammonium phosphate (100 kg ha^{-1}) was applied by broadcasting on the sowing date, and urea was applied in bands and fractionated as follows: 30% during pre-planting, 60% at the first irrigation, and 10% at the second irrigation. The irrigation water was 14 cm, and the irrigation time was 12 h. Harvest occurred 171 days after sowing (Figure 1).

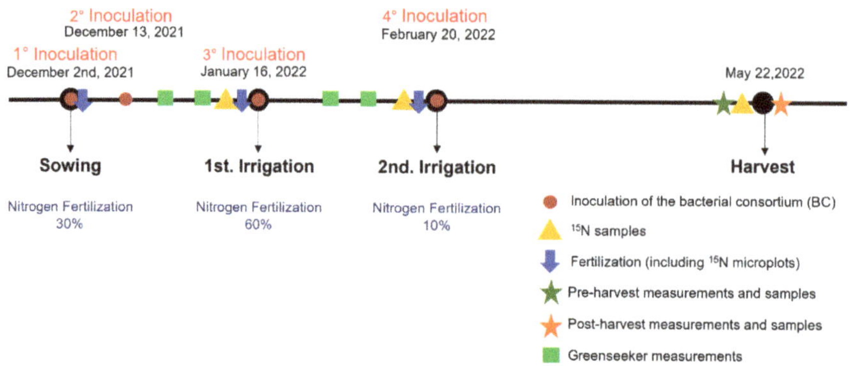

Figure 1. Timeline of the fieldwork developed during the wheat crop cycle.

2.2. Experimental Arrangement of the Field Experiment

The experimental design consisted of a split-plot (6.4 m × 10 m each plot) arrangement of six treatments [three (urea) fertilization doses (0, 120, and 240 kg N ha^{-1}) with and without the inoculation of the studied bacterial consortium (BC)], each replicated four times. Each plot contained an internal microplot (2.4 m × 2 m) exclusively designated for the application of ^{15}N-isotopically enriched urea at 1 atom % ^{15}N, following the rates of 120 and 240 kg N ha^{-1}. Urea of 5.09 atom % ^{15}N (Shanghai Research Institute of Chemical Industry Co., provided by the International Atomic Energy Agency), was isotopically diluted with conventional urea (with a natural abundance of ^{15}N) to achieve the desired enrichment of 1 atom % ^{15}N as stated in the equation described by Zapata and Axmann (1990) [21]. Urea with an enrichment of 1 atom % ^{15}N was then used in this experiment.

2.3. Bacterial Consortium

The native bacterial consortium consisted of *Bacillus cabrialesii* subsp. *cabrialesii* TE3^{T}, *Priestia megaterium* TRQ8, and *Bacillus paralicheniformis* TRQ65. These strains are currently cryopreserved at −80 °C in nutrient broth (NB) culture medium with glycerol (30%), at Colección de Microorganismos Edáficos y Endófitos Nativos (COLMENA, www.itson.mx/COLMENA, accessed on 24 October 2024).

Each bacterial strain was pre-cultured individually on Petri plates with Nutrient Agar (MCD Lab, Cat. 7141) and incubated at a temperature of 28 °C for 24 h. Each strain was also pre-inoculated in a minimal salt medium [in a volume of 20 mL, composed of glucose, 10 g L^{-1}; KH_2PO_4, 6.4 g L^{-1}; K_2HPO_4, 5.32 g L^{-1}; $(NH_4)_2SO_4$, 4 g L^{-1}; 0.4 g L^{-1} of $MgSO_4 \cdot H_2O$; 0.03 g L^{-1} of $FeSO_4 \cdot 7H_2O$; 0.044 g L^{-1} of $MnSO_4 \cdot H_2O$; and 0.021 g L^{-1} of $CaCl_2$], with a period of incubation of 24–48 h, at 180 rpm and a temperature of 28 °C, until an optical density (absorbance at the wavelength of 630 nm) of ~1.0 was reached. Later strains were inoculated in 500 mL of the same minimal salt culture medium individually, under the same conditions for 48 h, until reaching ≥1 × 10^{7} cell mL^{-1}.

The three bacterial strains (≥1 × 10^{7} cell mL^{-1}) were combined to form a consortium and were fermented in a BIOSTAT A bioreactor (Sartorius Stedim Biotech, Göttingen, Germany) in 5 L of a synthetic mineral medium composed of 20 g L^{-1} of glucose, 6.4 g L^{-1} of KH_2PO_4, 23.44 g L^{-1} of K_2HPO_4, 8 g L^{-1} of $(NH_4)_2SO_4$, 0.8 g L^{-1} of $MgSO_4 \cdot H_2O$, 0.06 g L^{-1} of $FeSO_4 \cdot 7H_2O$, 0.088 g L^{-1} of $MnSO_4 \cdot H_2O$, 0.126 g L^{-1} of $CaCl_2$, and $Na_3C_6H_5O$, 10.08 g L^{-1}. The bacterial consortium was fermented for 48 h, at the following conditions: temperature of 37 °C, pH of 7.0, agitation of 400 rpm, with constant air flow of 0.75 vvm, and silicon antifoam addition (30% p/v).

The number of viable bacterial cells was quantified as colony-forming units per milliliter (CFU mL^{-1}). Once a density of 1×10^{10} CFU mL^{-1} was obtained, the inoculant was applied to the crop (300 mL by each row of 0.8 m × 10 m long), manually directly to the soil (1×10^9 cells $plant^{-1}$) at the sowing date; 11 days after sowing (wheat emergence); 45 days after sowing (tillering stage), hours before the first irrigation; and 80 days after sowing (milk stage of development) before the second irrigation (Figure 1).

2.4. Measurement of Yield and Quality Parameters

The measurements of spikes m^{-2} (n = 12 per treatment), spike length (n = 36 per treatment), and the number of grains per spike (n = 62 per treatment) were registered at physiological maturity (pre-harvest). At the harvest, discarding the furrows of the edges, wheat biomass was collected manually using sickles, from the center of each plot. Every sheaf of wheat (per plot/replicate) was weighed on a scale (to estimate the total aboveground biomass), and then their grains were weighed after using a wheat threshing machine (to record grain weight per sample (n = 12 per treatment), considering the percentage of moisture at harvest, which was 8.8%). Then, straw yield and grain yield (adjusted to 14% moisture) were estimated, extrapolating each weight per area to tons per hectare. The harvest index was calculated by dividing grain yield by above-ground biomass (n = 12 per treatment). Hectoliter weight (n = 12) and 1000-grain weight of grains samples of each plot (n = 12) were estimated with a SeedCount Digital Image Analysis System SC5000 (Next Instruments, Sydney, Australia). On the other hand, a DA 7200 NIR analysis system (Perten Instruments, Sweden) was used to measure the percentage of grain (n = 12) protein at 12.5% humidity by near-infrared spectroscopy, which was calibrated based on official methods AACC 39-10.01 and 45-15.01 [22]. Additionally, wholemeal yellowness (Minolta b* value) was estimated utilizing a colorimeter (Konica Minolta, Tokyo, Japan) (n = 12); the SDS sedimentation index (n = 12) was calculated by dividing the SDS sedimentation volume by the sample protein content; sodium dodecyl sulfate (SDS) sedimentation volume (n = 12) was determined according to the method of Peña et al. (1990) [23].

2.5. ^{15}N-Nitrogen-Use Efficiency

To estimate ^{15}N-nitrogen-use efficiency with isotopic techniques of plant, straw, grain, and soil, samples were collected from the ^{15}N microplots (from the central furrow: an internal area of 0.8 m × 1 m, after excluding the edges) [21]. The collection of plant samples was before each fertilization event (45 and 80 days after sowing, in the tillering stage and the milk stage of development, respectively). Straw, grain, and soil samples were collected at harvest from the same internal sampling area of the microplot. The soil was collected at 30 cm depth, dried at room temperature, and sifted with a No. 100 mesh sieve (to achieve a particle size of 0.149 mm). Plant, straw, and grain samples were milled to obtain smaller sample sizes (straw and plant particle size <2 mm, grain particle size <0.3 mm).

All ^{15}N-enriched samples were processed in the Department of Plant Sciences of the University of California, at the Stable Isotope Facility (SIF) in Davis, with a continuous flow Elemental Analyzer coupled with an Isotope Ratio Mass Spectrometer (EA-IRMS). The equipment determined the percentage of ^{15}N atoms on all samples (plant, soil, straw, and grain; a ^{15}N-enriched urea sample was used as a control). Then, the NUE in the samples was calculated according to Zapata and Axmann (1990) [21], obtaining the nitrogen derived from the soil (Ndfs), subtracting the nitrogen derived from the fertilizer (Ndff), from the total N in the sample.

2.6. *In Vitro Assay of the Microbial Consortium Response to Different Rates of N and C (Ammonium Sulfate and Glucose)*

To determine the influence of varying carbon and nitrogen concentrations in the culture medium on the growth of the consortium strains, nine treatments comprising three replicates of three distinct N (ammonium sulfate) and C (glucose) concentrations were established in Elisa plates. Each strain of the bacterial consortium (*Bacillus cabrialesii* subsp. *cabrialesii* TE3^T, *Priestia megaterium* TRQ8, and *Bacillus paralicheniformis* TRQ65) was inoculated and fermented in individual centrifuge tubes with 15 mL of a synthetic mineral medium, of composition previously mentioned in the preparation of the bacterial consortium for the field application [glucose, 10 g L^{-1}; KH_2PO_4, 6.4 g L^{-1}; K_2HPO_4, 5.32 g L^{-1}; $(NH_4)_2$ SO_4, 4 g L^{-1}; 0.4 g L^{-1} of $MgSO_4{\cdot}H_2O$; 0.03 g L^{-1} of $FeSO_4{\cdot}7H_2O$; 0.044 g L^{-1} of $MnSO_4{\cdot}H_2O$; and 0.021 g L^{-1} of $CaCl_2$], at 30 °C and 180 rpm, for ~24 h, until reaching an optical density of ~1.0 (630 nm). Then, they were co-inoculated and co-fermented in a flask with 1 L of the synthetic mineral medium at the same conditions as the preinocula. After the optical density reached ~1.0, 0.1 mL of the inoculum was added to each well of the Elisa plates.

The base conditions of N and C were the original synthetic mineral medium, which was developed and selected due to the observed optimal growth of the microorganisms in the culture medium in those conditions. A total volume of 1 mL of the mineral medium was added to each well to establish the treatments. Later, proportional amounts of glucose and ammonium sulfate were added (Table 2) to form the treatments (3 replicates each one) that were (i) low N + low C; (ii) low N + medium C; (iii) low N + high C; (iv) medium N + low C; (v) medium N + medium C; (vi) medium N + high C; (vii) high N + low C; (viii) high N + medium C; and (ix) high N + high C.

Table 2. Necessary concentrations of ammonium sulfate and glucose to establish each treatment.

Component	N or C Content in the Treatment	Desired Composition (g/L)	Additional Mass to Make Up the Treatment (g/L)	Additional Mass to Make Up the Treatment in Each Well of 1 mL (g/1.0 mL)
$(NH_4)_2SO_4$	Low N	4	0	0
$(NH_4)_2SO_4$	Medium N	9.6000	5.6000	0.0056
$(NH_4)_2SO_4$	High N	19.2000	15.2000	0.0152
$C_6H_{12}O_8$	Low C	10	0	0
$C_6H_{12}O_8$	Medium C	20	10	0.0100
$C_6H_{12}O_8$	High C	27.7778	17.7778	0.0178

The optical density of all treatments was measured at 630 nm each hour for 21 h. At 14 h, when all treatments reached a value of absorbance of around 1.0, a serial dilution technique was used (0.1 mL of the sample in 0.9 mL of sterile water) to count the number of cells and spores (10^{-4} and 10^{-5}) and to count only spores (10^{-1} and 10^{-2}, raising the temperature to 80 °C for 15 min). Then, 0.1 mL of each dilution was added to Petri dishes with nutrient agar (5 g L^{-1} of peptone, 3 g L^{-1} yeast extract, and 15 g L^{-1} of bacteriological agar) and incubated for 24 h and 37 °C. The colony-forming units per milliliter (CFU mL^{-1}) were quantified for each treatment.

2.7. *Statistical Analysis*

Data components were processed with a two-way analysis of variance (ANOVA) in InfoStat, considering the nitrogen level (kg N ha^{-1} of nitrogen applied), the bacterial consortium, and their interactions as mixed factors. All factors were normally distributed (Shapiro–Wilk test). Levene's test of homogeneity of variances indicated that there are no significant differences between the variances of the treatments. Statistical results were validated with a significance level of $p < 0.05$ and $p < 0.01$. Differences between the parameters' means were assessed using Duncan's Multiple Range Test at a 95% confidence level.

3. Results and Discussion

The inoculation of the BC caused a significant increase ($p < 0.05$) in plant height on treatments of 0 and 240 kg N ha^{-1} vs. the uninoculated ones at the same N rate (Supplementary Table S2). Plant height was positively impacted by the inoculation of the bacterial consortium, as well as by the reduction in N fertilization to 0 kg N ha^{-1} (Table 3). Similarly, a remarkable tendency of higher values of root depth in all the inoculated treatments vs. uninoculated was observed (from 12.1 cm more) (Table 3). These positive outcomes could be due to the previously observed positive effects of the studied BC, for example, indole-3-acetic acid (IAA) biosynthesis, which is involved in several physiological processes, such as cell elongation and division, tissue differentiation, germination stimulation, influences root development, photosynthesis, pigment formation, biosynthesis of metabolites, and resistance to stress conditions [24]. In addition, strains contained in the studied BC can solubilize phosphate into a more accessible form to be absorbed by wheat plants. Phosphorus, as a macronutrient for crops, is the second most important mineral, an elemental component of nucleic acids, nucleotides, phospholipids enzymes, and coenzymes. Thus, it is involved in numerous physiological and biochemical reactions, such as photosynthesis, root and stem development, formation of flowers and seeds, crop maturation, and resistance to some diseases [25,26]. Furthermore, Valenzuela-Ruiz et al. (2019) [16] demonstrated significant increments in shoot height (26%), root length (36%), shoot dry weight (100%), stem diameter (53%), and biovolume index (146%) when strain TRQ65 was inoculated in wheat (vs. uninoculated wheat plants). Also, *P. megaterium* TRQ8, *B. cabrialesii* subsp. *cabrialesii* TE3^T, and *B. paralicheniformis* TRQ65 share more than 50% of biofertilization-related genes associated with CO_2 fixation, N and Fe acquisition, phosphate and K solubilization, and P assimilation, as well as genes related to the production of siderophores and stress response [5].

The number of spikes per square meter in all inoculated treatments resulted in optimal values for high wheat yield (200–300 spikes m^{-2}), without significant differences among them (Supplementary Table S2) [27]. However, a significant and positive effect of the inoculation of the BC was observed in this study (25 more spikes per square meter) compared to the uninoculated ones (Table 3); for example, a significant increase ($p < 0.05$) of 16.7% in spikes m^{-2} was observed in the 120 kg N ha^{-1} treatment when the CB was inoculated (Supplementary Table S2). Also, not applying N fertilization caused a positive impact on this trait, compared to 120 and 240 kg N ha^{-1} applied (Table 3). In the case of grains per spike, this parameter increased with the highest doses of N applied (55 grains per spike, vs. 51 and 52, Table 3).

In this sense, Oksel et al. (2022) [28] reported that the individual inoculation of seven *Bacillus* strains on wheat in Turkey, compared to an uninoculated control, the inoculation led to an increase in biometric traits; however, the inoculation of the *Bacillus* strains did not show significant increments in the number of grains per spike, suggesting that the positive effect of the inoculation of these strain on wheat yield could be through regulating several physiological traits.

Table 3. Wheat quantitative traits of yield and harvest indexes.

Factor		Plant Height (cm)		Root Depth (cm)		Spikes m^{-2}		Grains $Spike^{-1}$		Spike Size (cm)		Grain Yield (ton ha^{-1})		Straw Yield (ton ha^{-1})		Harvest Index	
		Mean	Standard Error	Mean	Standard Error	Mean	Standard Error	Mean	Standard Error	Mean	Standard Error	Mean	Standard Error	Mean	Standard Error	Mean	Standard Error
N Level, kg N ha^{-1} (A)	0 kg N ha^{-1}	79.7 b	0.25	108.3 a	5.61	274 b	6.1	51 a	0.61	6.3 a	0.06	5.01 a	0.14	3.57 a	0.12	0.59 a	0.0032
	120 kg N ha^{-1}	78.2 a	0.25	106.8 a	5.61	239 a	6.0	52 a	0.60	6.4 a	0.06	6.33 b	0.12	3.71 a	0.11	0.61 b	0.0038
	240 kg N ha^{-1}	78.0 a	0.25	96.3 a	5.61	256 a	6.0	55 b	0.58	6.9 b	0.06	6.77 c	0.12	4.22 b	0.10	0.63 c	0.0033
Bacterial consortium (B)	Uninoculated	77.8 a	0.21	97.8 a	4.58	244 a	5.0	53 a	0.49	6.5 a	0.05	6.18 a	0.10	4.01 b	0.08	0.61 a	0.0027
	Inoculated	79.4 b	0.21	109.7 a	4.58	269 b	4.8	53 a	0.48	6.5 a	0.05	5.90 a	0.11	3.66 a	0.09	0.61 a	0.0029
A × B	F value	17.83		0.24		0.91		4.79		21.24		14		9.63		3.25	
	p-value	<0.0001 **		0.7932		0.4095		0.0089 **		<0.0001 **		<0.0001 **		0.0004 **		0.0456 *	

Values in the same column indicate the mean, and different letters indicate a significant difference between treatments of the same parameter, using the Duncan test ($p < 0.05$). The bacterial consortium was composed of *Bacillus cabrialesii* subsp. *cabrialesii* TE3^T, *Priestia megaterium* TRQ8, and *Bacillus paralicheniformis* TRQ65. Asterisks indicate the significance level by the Duncan test [$p < 0.05$ (*)], $p < 0.01$ (**)].

In our essays, spike size significantly increased by 9% ($p < 0.05$) on the 240 kg N ha^{-1} + CB treatment compared to the uninoculated treatment at the same rate of N (7.2 vs. 6.6 cm, Supplementary Table S2), but in the unfertilized treatments (0 kg N ha^{-1}) a decrease of 7.7% was observed in the inoculated vs. uninoculated treatment (6 vs. 6.5 cm). In this sense, this trait was more influenced by the application of higher doses of N (240 kg N ha^{-1}), than by the inoculation of the BC (Table 3).

Regarding grain yield, a significant ($p < 0.05$) increase due to the inoculation of the BC was observed at 240 kg N ha^{-1} (7.06 vs. 6.48 tons ha^{-1}). On the other hand, a decrease of 1.26 ton ha^{-1} was observed by the inoculation of BC vs. the uninoculated treatment under 0 kg N ha^{-1} (Supplementary Table S2). In general, in this crop cycle, the doses of N had a greater impact on grain yield than the inoculation of the bacterial consortium (Table 3). This effect of inoculation of the BC on wheat contrasts with that observed in previous field experiments carried out by our group, under similar field conditions. Ayala-Zepeda et al. (2024) [5] reported that the inoculation of strains TRQ8, TRQ65, and TE3^{T} to durum wheat resulted in a yield increase of 1.1 ton ha^{-1} under 0 kg N ha^{-1} (123 kg N ha^{-1} as residual in the soil) compared with its uninoculated control, and an increment of 2.0 ton ha^{-1} under 120 kg N ha^{-1} (104 kg N ha^{-1} a residual in the soil) compared with uninoculated treatment. Our essay was carried out at the same location, agricultural practices, and N applied, as well as similar soil physical properties and climatic conditions compared with Ayala-Zepeda et al. (2024). However, in the present study, the N content in the soil on the sowing date was 183 kg N ha^{-1}, and the N was fractionated differently at 30, 60, and 10%, in comparison with 33%, 33%, and 33% used by Ayala-Zepeda et al. (2024) [5].

Another distinct finding was reported by Ibarra-Villarreal et al. (2023) [19], who inoculated a bacterial consortium (*Bacillus subtilis* TSO9, *Bacillus cabrialesii* subsp. *tritici* TSO2^{T}, *Bacillus subtilis* TSO22, *Bacillus paralicheniformis* TRQ65, and *Priestia megaterium* TRQ8), to durum wheat in the same location, under the same agricultural practices, three different N rates applied, and similar soil and climatic conditions, obtaining a increment of 1 ton ha^{-1} in grain yield in the inoculated treatment under 130 kg N ha^{-1} (medium N fertilization) compared to the uninoculated control.

Various studies have shown increases in crop productivity, yield parameters, or nitrogen-use efficiency with the inoculation of plant growth-promoting bacteria under medium doses of N fertilization. Gaspareto et al. 2023 [29] tested the co-inoculation of winter wheat with *Azospirillum brasilense* and *Bacillus subtilis* under five N doses (0, 40, 80, 120, and 160 kg ha^{-1}, applied from urea) in Brazilian Cerrado. They suggest that N fertilization can be reduced by application of those strains under a no-tillage system; their results show that a medium rate of fertilization of 80 kg ha^{-1}, nitrogen-use efficiency, grain N accumulation, and number of grains spikes^{-1} were significantly increased. On the other hand, Pardo-Díaz et al. (2021) [30] found that the effect of PGPB inoculation on plant growth was observed only with 50% of the N fertilization dose. Moreover, Saia et al. (2015) [31] found that durum wheat inoculation with arbuscular mycorrhizal fungi (AMF) and plant growth-promoting rhizobacteria (PGPR) (alone or in combination) improved plant growth, and N uptake, in comparison with uninoculated controls.

According to these results, regarding the use of bacterial inoculants as a sustainable alternative to wheat production, not only is the nitrogen fertilization dose important, but so is fractionation or the N split application (the total N at specific stages of wheat growth). Moreover, in earlier works, we proposed that positive effects on wheat yield in the Yaqui Valley inoculated with the bacterial consortium used in this study are most likely to be observed with total N concentrations of 123–225 kg N ha^{-1}, considering soil N content and N from fertilization (but with a different N split application: 33% at the sowing date, 33% with the first irrigation event, and 33% with the second irrigation) [5]. In this sense, in this crop cycle, the treatments of 120 and 240 kg N ha^{-1} applied surpass this range (adding 183 kg N ha^{-1} of the soil).

It has been observed that as N levels surpass 225 kg N ha^{-1}, wheat yield and productivity begin to decrease, which may be due to exceeded crop nutrient requirements, as well as adverse effects on the photosynthesis rate and grain filling potential in cereal crops [32]. Thus, high doses of N in the soil, along with low efficiency, can cause toxicity, salinity, and pollution of soils and water bodies [33]. Also, it has been reported that N fertilizer input in the short term altered the soil NO_3^- contents, consequently impacting negatively the predominant rhizosphere soil bacterial species [34].

In agricultural soils, high doses of N affect growth and some beneficial effects of plant growth-promoting microorganisms [13] and decrease microbial functional diversity [35]. In general, fertilization changes the activity and community structure of soil microorganisms [34,36]. In this sense, results of varying carbon and nitrogen in vitro essay (and the inoculation of the BC that was used in the field), showed that the highest sporulation occurred with the highest nitrogen content in the medium, and with the three different C contents (4.90×10^4 UFC mL^{-1}, 1.07×10^5 UFC mL^{-1}, 4.80×10^4 UFC mL^{-1}: low, medium, and high, respectively) (Figure 2). Although this dormant state (endospore formation) is a desirable trait for the development of cost-effective microbial inoculants, microorganisms also produce spores as a survival mechanism to adverse environmental conditions and different types of stress, resulting in a metabolically inactive state [37].

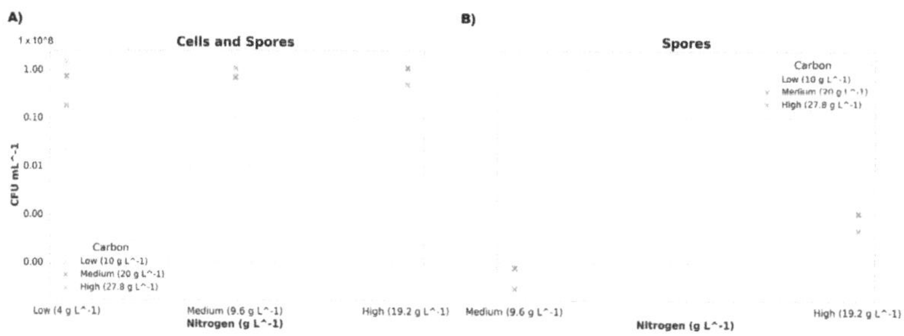

Figure 2. The growth response of the microbial consortium to varying rates of nitrogen and carbon, as detailed in Table 2: (**A**) Quantification of colony forming units (CFU) per milliliter, including both cells and spores. (**B**) Quantification of colony-forming Units (CFU) per milliliter of spores following a 15 min thermal shock treatment.

The above could be a factor limiting the BC beneficial effect at high N doses, considering that there are certain important growth stages of wheat where fertilization and the application of the BC are synchronized and where plant growth-promoting activity is crucial to observe positive impacts on plant biometric parameters, yield, and quality. These results suggest that these strains could be forming endospores when high quantities of N are applied to the soil, instead of being in an active state at those specific and critical moments (sowing date, tillering stage, booting stage, among others).

Thus, medium rates of nitrogen are required for adequate BC growth, since the consortium strains resulted in 7.40×10^7 UFC mL^{-1} (with low C), 6.90×10^7 UFC mL^{-1} (with medium C), and 1.06×10^8 UFC mL^{-1} (along with high C) cells and spores; also, in 2.80×10^3 UFC mL^{-1} of spores with high C content, and 7.60×10^3 UFC mL^{-1} of spores with medium C content (Figure 2).

Other studies have shown how high N fertilization rates affect soil microbial flora. Perez et al. [38] studied the influence of nitrogen in soil microbial population of a *Coffea canephora* Pierre cultivation and found that the application of 90 and 135 kg N ha^{-1} resulted in the highest values of bacteria populations (116–139 $\times 10^7$ CFUg^{-1}); 150 kg N ha^{-1}

reduced the populations to their initial state (107×10^7 CFUg^{-1}); and the highest rate of fertilization, 200 kg N ha^{-1}, depleted the populations of bacteria to 50×10^7 CFUg^{-1}. In comparison with Pardo-Díaz et al. [30], who evaluated *Herbaspirillum* sp. AP21 and *Azospirillum brasilense* D7 (without inoculation, individual, and in combination) with 50% and 75% of urea, vs. controls without PGPB with 0% and 100% of urea on a Perennial ryegrass (*Lolium perenne* L.) var. One 50 and red clover (*Trifolium pratense* L.) intercropping system, they reported that the inoculation of the PGPB decreased the rhizosphere bacterial diversity, but nitrogen doses did not cause significant changes in the bacterial community.

Regarding the quality of the grain, the traits of hectoliter weight and protein percentage (Table 4) were positively ($p < 0.05$) impacted by the increased doses of applied nitrogen dose, more than the inoculation of the studied bacterial strains (Table 4). Similarly, Lerner et al. [39] reported that the protein content in grains of different Argentine varieties of bread wheat was highly dependent on the level of N.

Concerning the trait of 1000-grain weight, no significant difference was observed among treatments due to the inoculation of the BC (Table 4), and it has been stated that increases in spikes per m^2, 1000-grain weight, and number of grains per spike are necessary to achieve an increment of wheat yield [40]. This indicator is consistent with the yields obtained (except for the lowest two values when 0 kg N ha^{-1} was applied) (Supplementary Tables S1 and S2). The range between all results of Minolta b* values of yellowness was 14.20 to 14.73 (Supplementary Table S3), without a significant effect due to the inoculation of the BC, and a tendency of higher values with increasing N rates (Table 4). These results were coherent, as this quality trait is reported to be more stable in different conditions and mainly genetically determined [41]. All values of yellowness in this study met the quality requirements for this wheat variety [42].

However, a low protein percentage in general (among all treatments) was observed, which is the main trait considered to define an appropriate wheat grain quality for industrial purposes in the region. This effect could be due to low N availability in advanced stages of growth (e.g., heading or grain filling) and N was not available for a beneficial impact on quality (protein), considering that the last fractionation of N was 10% in the grain filling stage. It is important to highlight that wheat gluten is composed of proteins divided into gliadins (responsible for dough extensibility and viscosity) and glutenins (main determinants of dough strength), which must be in balance to achieve desirable quality characteristics. Therefore, it has been reported that the γ-gliadin gene family is largely regulated by the N supply in the course of wheat grain development [43,44]. Otherwise, Lerner et al. (2016) [39] reported that the relationship between dough extensibility and recovery efficiency of N in grain was more related to the partition of N to harvestable destinations than to the absorption of N from the soil.

In summary, the inoculation of the bacterial consortium had no significant effect on the studied quality parameters, which were more stable and/or influenced by different N doses. Although beneficial bacteria have the potential to improve plant growth and fertilizer uptake, excess nitrogen can induce sporulation and limit their metabolic activity at critical stages of crop development. This highlights the importance of properly managing nitrogen doses to maximize the benefits of microbial inoculants in crops.

In this study, it was demonstrated that fertilization of 120 kg of N ha^{-1} is more efficient (NUE: 44.32%) than the full fertilization dose of 240 kg of N ha^{-1} (NUE: 27.51%) (Table 5), due to a lower amount of N fertilizer applied. Although the N derived from the fertilizer at physiological maturity (Figure 3) was quantitatively higher in the 240 kg N ha^{-1} treatments (69.1 ± 30.3 kg ha^{-1} and 62.9 ± 11.5 kg N ha^{-1}) compared to the 120 kg of N ha^{-1} treatments (52.4 ± 7.6 kg ha^{-1} and 54.0 ± 23.2 kg ha^{-1}), the difference was not statistically significant.

Table 4. Wheat quality traits of the studied treatments.

Factor		Hectoliter Weight (kg hL^{-1})		1000 Grain Weight (g)		Protein (%)		SDS Sedimentation Test (mL)		SDS/Protein Index		Wholemeal Yellowness (Minolta, b)	
		Mean	Standard Error	Mean	Standard Error	Mean	Standard Error	Mean	Standard Error	Mean	Standard Error	Mean	Standard Error
N Level, kg N ha^{-1} (A)	0 kg N ha^{-1}	78.85 a	0.13	57.94 b	0.42	9.20 a	0.16	11.31 a	0.17	1.25 b	0.02	14.23 a	0.06
	120 kg N ha^{-1}	80.89 b	0.13	58.08 b	0.42	11.09 b	0.16	11.29 a	0.16	1.02 a	0.02	14.52 b	0.06
	240 kg N ha^{-1}	80.76 b	0.13	56.61 a	0.42	11.81 c	0.16	12.16 b	0.16	1.04 a	0.02	14.68 b	0.06
Bacterial consortium (B)	Uninoculated	80.48 a	0.11	57.69 a	0.35	10.58 a	0.13	11.49 a	0.14	1.10 a	0.01	14.53 a	0.05
	Inoculated	80.52 a	0.11	57.40 a	0.35	10.82 a	0.13	11.68 a	0.14	1.10 a	0.01	14.43 a	0.05
A × B	F value	0.72		3.03		0.8		0.87		2.33		0.12	
	p-value	0.4911		0.0553		0.4538		0.4228		0.1056		0.8851	

Values in the same column indicate the mean, and different letters indicate a significant difference between treatments of the same parameter, using the Duncan test ($p < 0.05$). The bacterial consortium was composed of *Bacillus cabrialesii* subsp. *cabrialesii* TE3[T], *Priestia megaterium* TRQ8, and *Bacillus paralicheniformis* TRQ65.

Table 5. Nitrogen (^{15}N) uses the efficiency of wheat under two different doses of urea and the inoculation of a native bacterial consortium.

Factor		Nitrogen-Use Efficiency of ^{15}N Labeled Urea (%) Grain (%)		Straw (%)		Whole Plant (%)		^{15}N Remnant in the First 30 cm of Soil (%)		Losses of ^{15}N (%)	
		Mean	Standard Error	Mean	Standard Error	Mean	Standard Error	Mean	Standard Error	Mean	Standard Error
N Level, kg N ha^{-1} (A)	120 kg N ha^{-1}	40.53 b	4.08	3.79 a	0.33	44.32 b	4.30	4.88 a	0.84	47.22 a	3.77
	240 kg N ha^{-1}	24.48 a	4.41	3.03 a	0.35	27.51 a	4.64	6.39 a	0.78	65.85 b	3.77
Bacterial consortium (B)	Uninoculated	33.29 a	4.41	2.95 a	0.35	36.23 a	4.64	4.17 a	0.78	59.33 a	3.77
	Inoculated	31.73 a	4.08	3.87 a	0.33	35.60 a	4.30	7.10 b	0.84	53.73 a	3.77
A × B	F value	0.04		2.75		0.09		9.89		0.17	
	p-value	0.8526		0.1257		0.7652		0.0093 **		0.6931	

One atom % ^{15}N excess labeled urea was used in this experiment. The bacterial consortium was composed of *Bacillus cabrialesii* subsp. *cabrialesii* TE3[T], *Priestia megaterium* TRQ8, and *Bacillus paralicheniformis* TRQ65. Different letters in the same column indicate a significant difference using the Duncan test ($p < 0.05$). Asterisks indicate the significance level by the Duncan test [$p < 0.01$ (**)].

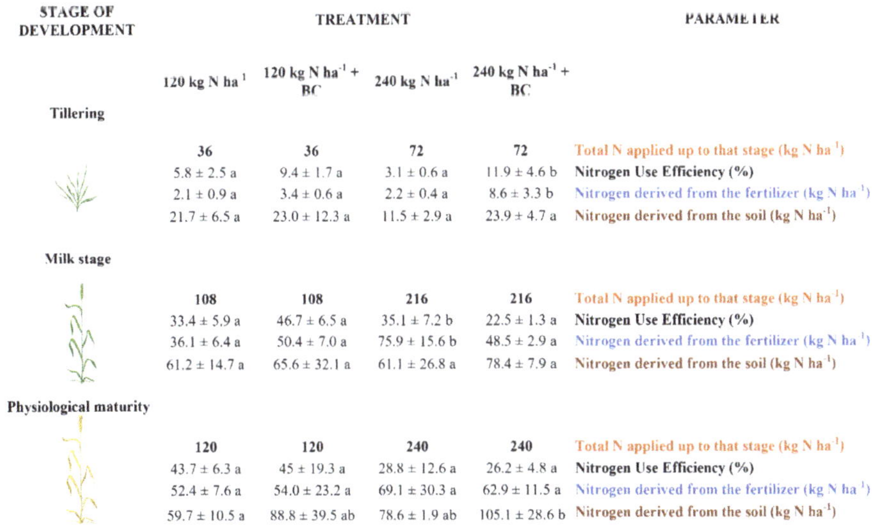

STAGE OF DEVELOPMENT	TREATMENT				PARAMETER
	120 kg N ha^{-1}	120 kg N ha^{-1} + BC	240 kg N ha^{-1}	240 kg N ha^{-1} + BC	
Tillering	36	36	72	72	Total N applied up to that stage (kg N ha^{-1})
	5.8 ± 2.5 a	9.4 ± 1.7 a	3.1 ± 0.6 a	11.9 ± 4.6 b	Nitrogen Use Efficiency (%)
	2.1 ± 0.9 a	3.4 ± 0.6 a	2.2 ± 0.4 a	8.6 ± 3.3 b	Nitrogen derived from the fertilizer (kg N ha^{-1})
	21.7 ± 6.5 a	23.0 ± 12.3 a	11.5 ± 2.9 a	23.9 ± 4.7 a	Nitrogen derived from the soil (kg N ha^{-1})
Milk stage	108	108	216	216	Total N applied up to that stage (kg N ha^{-1})
	33.4 ± 5.9 a	46.7 ± 6.5 a	35.1 ± 7.2 b	22.5 ± 1.3 a	Nitrogen Use Efficiency (%)
	36.1 ± 6.4 a	50.4 ± 7.0 a	75.9 ± 15.6 b	48.5 ± 2.9 a	Nitrogen derived from the fertilizer (kg N ha^{-1})
	61.2 ± 14.7 a	65.6 ± 32.1 a	61.1 ± 26.8 a	78.4 ± 7.9 a	Nitrogen derived from the soil (kg N ha^{-1})
Physiological maturity	120	120	240	240	Total N applied up to that stage (kg N ha^{-1})
	43.7 ± 6.3 a	45 ± 19.3 a	28.8 ± 12.6 a	26.2 ± 4.8 a	Nitrogen Use Efficiency (%)
	52.4 ± 7.6 a	54.0 ± 23.2 a	69.1 ± 30.3 a	62.9 ± 11.5 a	Nitrogen derived from the fertilizer (kg N ha^{-1})
	59.7 ± 10.5 a	88.8 ± 39.5 ab	78.6 ± 1.9 ab	105.1 ± 28.6 b	Nitrogen derived from the soil (kg N ha^{-1})

Figure 3. Effect of nitrogen dose and the inoculation of a bacterial consortium on nitrogen-use efficiency (^{15}N) at different durum wheat growth stages. 1 atom % ^{15}N excess labeled urea was applied at planting (30%), 45 days after sowing (60%), and 80 days after sowing (10%) at two different rates of N, with and without the inoculation of a bacterial consortium (BC) formed by *Bacillus cabrialesii* subsp. *cabrialesii* TE3^{T}, *Priestia megaterium* TRQ8, and *Bacillus paralicheniformis* TRQ65. 'Total N applied up to that stage' is the sum of N applications per hectare before the date of the collection of that sample in that stage of development; 'Nitrogen-use Efficiency' is the percentage of the applied nitrogen fertilizer that was recovered by the crop. Different letters in the same row indicate a significant difference, using the Duncan test ($p < 0.05$).

Silveira et al. [45] studied the impact of three diazotrophic endophytic bacteria, IAC-AT-8 (*Azospirillum brasilense*), IAC-HT-11 (*Achromobacter insolitus*), and IAC-HT-12 (*Zoogloea ramigera*), on the metabolism, physiology, and growth of wheat (*Triticum aestivum hard* L.) plants under different levels of nitrogen urea: 0%, 50%, and 100% of the recommended N-fertilizer rate. They found increases in glutamine synthetase (GS) activity in 0% and 50% of the fertilizer (which also exhibited an increment in nitrate reductase—NR—activity) in plants colonized by *A. insolitus*. The GS activity is studied as an auxiliary parameter for the evaluation of N utilization; its increase is related to increments of N content and dry weight of shoot, root, and/or grain.

As previously mentioned, recent works by our group have demonstrated a positive effect of the BC on wheat yield at total N concentrations of 123–225 kg N ha^{-1} in the soil, maintaining good grain quality [5], emphasizing significant differences in inoculated treatments vs. their uninoculated controls (fractionating N: 33%, 33%, and 33% at each fertilization event). Also, increases in nitrogen-use efficiency (14 and 12.5%) at two consecutive agricultural cycles 2018–2019 and 2019–2020 were observed when the N fertilization dose was 120 kg N ha^{-1} (uninoculated and inoculated; NUE of 39.3 and 36.3% in 2018–2019; 46.9 and 42.8% in 2019–2020) vs. the total N rate fertilization of 240 kg N ha^{-1} (both uninoculated and inoculated; NUE of 25.3 and 26.3% in 2018–2019; 34.4 and 21% in 2019–2020) [5].

It is worth noting that fertilization at planting, measured at the tillering stage (Figure 3), showed a very low NUE, with values no higher than 11.9 ± 4.6%, and as low as 3.1 ± 0.6%. During this stage, more N was derived from the soil (from 11.5 ± 2.9% to 23.9 ± 4.7%). This effect could be attributed to the fertilization method, where urea is surface-applied in bands without irrigation at sowing, combined with root architecture that makes soil-derived N more accessible. These factors likely result in greater N losses from the fertilizer during early growth stages [46].

When high quantities of N are applied at early stages (with low plant uptake capacity), N losses from the soil–plant system are enhanced, which leads to low N recovery efficiency and low yields [32]. In this sense, it is important to note that in this study, when fertilizing with half the recommended N rate with the inoculation of the BC (120 kg N ha^{-1} + BC) significantly fewer nitrogen losses are observed (47.22%) compared to when wheat was fully fertilized (240 kg N ha^{-1}, 65.85%, Table 5 and Supplementary Table S4).

Moreover, it has been reported that the application of 20 to 40% less N (as compared to the recommended dose) can cause a significant increment in NUE of the wheat crop without affecting the yield, and it can reduce N losses [46]. However, at the tillering stage and physiological maturity (Figure 3), all inoculated treatments had a higher total N (from both soil and fertilizer), compared to the uninoculated controls at the same dose. By the milk stage, inoculation of the BC with 120 kg of N ha^{-1} also resulted in higher total N compared to the uninoculated treatment.

It has been reported that PGPB have a great impact on nitrogen use by increasing nitrate (NO_3^-) uptake capacity, whether it is by the direct stimulation of NO_3^- transport systems, or as a result of stimulated lateral root development [47].

Moreover, all treatments at the three studied physiological stages (except the treatment of 240 kg N ha^{-1}, uninoculated, at the milk stage) took more nitrogen from the soil than from the fertilizer (Figure 3), which represents another indicator of the importance of diminishing N fertilizer rates and/or to improve their management, showing, that the use of sustainable agricultural practices is essential.

It is important to highlight that the present investigation is not only about applying the PGPB but also about identifying the suitable conditions that would guarantee their maximum efficiency in the field. One of them is nitrogen management (the dose and fractionation of the applied N). Since there are no significant differences in some parameters due to the inoculation of BC (as the results observed in earlier works [5,19]), the main findings are that the same microorganisms depending on the dose and fractionation of N may or may not have the desired beneficial effect on wheat productivity.

Although conducted over a single agricultural cycle, this study lays the foundation for future research aimed at strengthening the analysis of the observed phenomena through different lines of research. Furthermore, these findings are crucial to refining recommendations on the BC application in conjunction with proper N management, offering farmers potential savings from fertilizer inputs or increased profits from higher wheat productivity in individual crop cycles, while promoting more sustainable practices.

Given that, to increase wheat productivity sustainably, two approaches are of importance: management practices (including crop rotation, cover crops, integrated pest management, N rate, N source, right moment of N fertilization, and N place of application, among others) and plant breeding (developing improved cultivars with desired traits such as tolerance to biotic and abiotic stresses, or high yield and quality, to mention a few). However, to guarantee the expression of genetic potential, most wheat breeding trials are carried out with high levels of N [48]. In this sense, it is recommended to continue testing crop cultivars with low N environments, along with sustainable alternatives such as the inoculation of native microorganisms, to explore their performance on N-limiting conditions, and to harness the potential of the strains, since it has been well documented that the beneficial effect of PGPM on crops is mostly presented with lower doses of N.

4. Conclusions

Bacterial inoculants based on PGPB represent a sustainable alternative to increase crop productivity and diminish the dependence on synthetic inputs, such as nitrogen fertilizers. However, their effectiveness depends not only on the target crop and the microbial inoculant characteristics but also on soil and environmental conditions, as well as management practices. In this regard, the inoculation of wheat with the bacterial consortium used in this study (*Bacillus cabrialesii* subsp. *cabrialesii* TE3^T, *Priestia megaterium* TRQ8, and *Bacillus paralicheniformis* TRQ65) had positive effects on the early growth stages, plant height, root depth, and spikes per square meter parameters (when the plants are fertilized at sowing with a fraction of 30% of the total N). However, no increases in grain yield due to the inoculation of the BC were observed when fertilization was reduced completely or by half (0 or 120 kg N ha^{-1}; as already observed in other crop cycles and studies conducted by our research group). This was probably due to the high residual N soil content (183 kg N ha^{-1}) along with the fractionation of N fertilization by 30%, 60%. and 10% (in sowing, stem elongation, and booting), due to the high N input when fertilized with 60% at stem elongation. Furthermore, the BC had no significant effect on the studied quality parameters, which were more stable and influenced by different N doses. Nevertheless, the inoculation of the bacterial consortium on wheat caused more N acquisition from both sources: soil and fertilizer, at different N fertilization rates and growth stages.

Fertilization had a significant effect on nitrogen-use efficiency (NUE) of the whole plant: 120 kg of N ha^{-1} was more efficient (NUE: 44.32%) than the full fertilization dose (NUE: 27.51%). Consequently, significantly greater losses of N from the soil-plant system are observed when wheat was fully fertilized (240 kg N ha^{-1}, 65.85%) compared to fertilization with half the recommended N rate with the BC inoculation (120 kg N ha^{-1} + BC, 47.22%).

Overall, when incorporating the studied bacterial inoculant into wheat crops, it is essential to use appropriate agricultural practices, including N management of no more than 225 kg N ha^{-1} (from both soil and fertilizer) and opt for a nitrogen fertilizer fractionation of 33%, 33%, and 33%. This short-term experiment lays the groundwork for developing more comprehensive and long-term studies that will elucidate effective management practices compatible with bacterial consortium inoculation, ultimately contributing to sustainable agriculture.

Supplementary Materials: The following supporting information can be downloaded at: https://www.mdpi.com/article/10.3390/app15031429/s1, Supplementary Table S1. Additional climatic conditions during the crop cycle; Supplementary Table S2. Wheat quantitative traits of yield and harvest indexes; Supplementary Table S3. Wheat quality traits of the studied treatments; Supplementary Table S4. Nitrogen (^{15}N) uses the efficiency of wheat under two different doses of urea and the inoculation of a native bacterial consortium.

Author Contributions: Investigation, formal analysis, data curation, writing—original draft preparation, visualization, M.A.-Z.; methodology, conceptualization, writing—review and editing, all authors; resources, S.d.l.S.-V. and M.I.I.; funding acquisition, conceptualization, supervision, project administration, validation, S.d.l.S.-V. All authors have read and agreed to the published version of the manuscript.

Funding: This research was supported by the Joint FAO/IAEA through the Regional Latin America Project RLA5077 "Enhancing Livelihood through Improving Water Use Efficiency Associated with Adaptation Strategies and Climate Change Mitigation in Agriculture (ARCAL CLVIII)".

Institutional Review Board Statement: Not applicable.

Informed Consent Statement: Not applicable.

Data Availability Statement: The raw data supporting the conclusions of this article will be made available by the authors on request.

Acknowledgments: The authors would like to thank Consejo Nacional de Humanidades, Ciencias y Tecnologías (CONAHCYT) for the doctorate scholarship CVU: 893199 (Marisol Ayala Zepeda); and to members of Laboratorio de Biotecnología del Recurso Microbiano for their support in the field labor, their revision, and feedback, especially thesis students of the RLA5077 project. The authors greatly acknowledge Brenda Valenzuela Aragon for the development of Figure 2.

Conflicts of Interest: The authors declare no conflicts of interest. The funders had no role in the design of the study; in the collection, analyses, or interpretation of data; in the writing of the manuscript; or in the decision to publish the results.

References

1. Khalid, A.; Hameed, A.; Tahir, M.F. Wheat Quality: A Review on Chemical Composition, Nutritional Attributes, Grain Anatomy, Types, Classification, and Function of Seed Storage Proteins in Bread Making Quality. *Front. Nutr.* **2023**, *10*, 1–14. [CrossRef] [PubMed]
2. Igrejas, G.; Ikeda, T.M.; Guzmán, C. *Wheat Quality for Improving Processing and Human Health*; Springer: Berlin/Heidelberg, Germany, 2020; ISBN 9783030341633.
3. SIAP, (Servicio de Información Agroalimentaria y Pesquera) Anuario Estadístico de La Producción Agrícola. Available online: https://nube.siap.gob.mx/cierreagricola/ (accessed on 1 May 2024).
4. Ortiz-Monasterio, J.I.; Raun, W. Paper Presented at International Workshop on Increasing Wheat Yield Potential, CIMMYT, Obregon, Mexico, 20–24 March 2006: Reduced Nitrogen and Improved Farm Income for Irrigated Spring Wheat in the Yaqui Valley, Mexico, Using Sensor Based Nitrogen Management. *J. Agric. Sci.* **2007**, *145*, 215–222. [CrossRef]
5. Ayala-Zepeda, M.; Valenzuela-Ruíz, V.; Parra-Cota, F.I.; Chinchilla-Soto, C.; de la Cruz-Torres, E.; Ibba, M.I.; Estrada-Alvarado, M.I.; de los Santos-Villalobos, S. Genomic Insights of a Native Bacterial Consortium for Wheat Production Sustainability. *Curr. Res. Microb. Sci.* **2024**, *6*, 100230. [CrossRef] [PubMed]
6. Sutton, M.A.; Howard, C.M.; Erisman, J.W.; Bealey, W.J.; Billen, G.; Bleeker, A.; Bouwman, A.F.; Grennfelt, P.; van Grinsven, H.; Grizzetti, B. The Challenge to Integrate Nitrogen Science and Policies: The European Nitrogen Assessment Approach. In *The European Nitrogen Assessment*; Cambridge University Press: Cambridge, UK, 2011; pp. 82–96. [CrossRef]
7. FIRA (Fideicomisos Instituidos en Relación con la Agricultura). *Actualización En Costos Publicados* . 2024. Available online: https://www.fira.gob.mx/InfEspDtoXML/TemasUsuario.jsp (accessed on 5 November 2023).
8. Han, M.; Okamoto, M.; Beatty, P.H.; Rothstein, S.J.; Good, A.G. The Genetics of Nitrogen Use Efficiency in Crop Plants. *Annu. Rev. Genet.* **2015**, *49*, 269–289. [CrossRef]
9. Ray, D.K.; Mueller, N.D.; West, P.C.; Foley, J.A. Yield Trends Are Insufficient to Double Global Crop Production by 2050. *PLoS ONE* **2013**, *8*, e66428. [CrossRef]
10. Fischer, T.; Ammar, K.; Monasterio, I.O.; Monjardino, M.; Singh, R.; Verhulst, N. Sixty Years of Irrigated Wheat Yield Increase in the Yaqui Valley of Mexico: Past Drivers, Prospects and Sustainability. *Field Crops Res.* **2022**, *283*, 108528. [CrossRef]
11. Aquino, J.P.A.; Antunes, J.E.L.; Bonifácio, A.; Rocha, S.M.B.; Amorim, M.R.; Alcântara Neto, F.; Araujo, A.S.F. Plant Growth-Promoting Bacteria Improve Growth and Nitrogen Metabolism in Maize and Sorghum. *Theor. Exp. Plant Physiol.* **2021**, *33*, 249–260. [CrossRef]
12. Thilagar, G.; Bagyaraj, D.J.; Raoca, M.S. Selected Microbial Consortia Developed for Chilly Reduces Application of Chemical Fertilizers by 50% Under Field Conditions. *Sci. Hortic.* **2016**, *198*, 27–35. [CrossRef]
13. Glick, B.R. Plant Growth-Promoting Bacteria: Mechanisms and Applications. *Scientifica (Cairo)* **2012**, *2012*, 15. [CrossRef]
14. Chami, B.; Touzout, N.; Guemouri-Athmani, S.; Baali-Cherif, D.; Mihoub, A.; Černý, J.; Saeed, M.F.; Jamal, A.; Yassine, H.M.; Dewir, Y.H.; et al. Bacterial Inoculation and Co-Inoculation Improves Durum Wheat Productivity in Alkaline Calcareous Soils. *Phyton-Int. J. Exp. Bot.* **2024**, *93*, 3313–3329. [CrossRef]
15. De los Santos-Villalobos, S.; Robles, R.I.; Parra-Cota, F.I.; Larsen, J.; Lozano, P.; Tiedje, J.M. *Bacillus cabrialesii* sp. Nov., an Endophytic Plant Growth Promoting Bacterium Isolated from Wheat (*Triticum turgidum* subsp. Durum) in the Yaqui Valley, Mexico. *Int. J. Syst. Evol. Microbiol.* **2019**, *69*, 3939–3945. [CrossRef] [PubMed]
16. Valenzuela-Ruiz, V.; Robles-Montoya, R.I.; Parra-Cota, F.I.; Santoyo, G.; del Carmen Orozco-Mosqueda, M.; Rodríguez-Ramírez, R.; de los Santos-Villalobos, S. Draft Genome Sequence of Bacillus Paralicheniformis TRQ65, a Biological Control Agent and Plant Growth-Promoting Bacterium Isolated from Wheat (*Triticum turgidum* subsp. Durum) Rhizosphere in the Yaqui Valley, Mexico. *3 Biotech* **2019**, *9*, 436. [CrossRef] [PubMed]

17. Rojas Padilla, J.; Chaparro Encinas, L.A.; Robles Montoya, R.I.; De los Santos Villalobos, S. Promoción de Crecimiento En Trigo (*Triticum turgidum* L. subsp. Durum) Por La Co-Inoculación de Cepas Nativas de Bacillus Aisladas Del Valle Del Yaqui, México. *Nova Sci.* **2020**, *12*, 1–27. [CrossRef]
18. Rojas-Padilla, J.; de-Bashan, L.E.; Parra-Cota, F.I.; Rocha-Estrada, J.; de los Santos-Villalobos, S. Microencapsulation of Bacillus Strains for Improving Wheat (*Triticum turgidum* subsp. Durum) Growth and Development. *Plants* **2022**, *11*, 2920. [CrossRef]
19. Ibarra-Villarreal, A.L.; Villarreal-Delgado, M.F.; Parra-Cota, F.I.; Yepez, E.A.; Guzmán, C.; Gutierrez-Coronado, M.A.; Valdez, L.C.; Saint-Pierre, C.; de Los Santos-Villalobos, S. Effect of a Native Bacterial Consortium on Growth, Yield, and Grain Quality of Durum Wheat (*Triticum turgidum* L. subsp. Durum) Under Different Nitrogen Rates in the Yaqui Valley, Mexico. *Plant Signal Behav.* **2023**, *18*, 2219837. [CrossRef]
20. González-Figueroa, S.S.; Vera-Núñez, J.A.; Peña-Cabriales, J.J.; Báez-Pérez, A.; Grageda-Cabrera, O.A. Uso Eficiente De Nitrógeno En Aplicaciones Fraccionadas De Fertilizante Marcado Con 15N En Trigo. *Rev. Fitotec. Mex.* **2022**, *45*, 437. [CrossRef]
21. Zapata, F.; Axmann, H. Stable and Radioactive Isotopes. In *Use of Nuclear Techniques in Studies of Soil-Plant Relationships*; Hardarson, G., Ed.; International Atomic Energy Agency (IAEA): Vienna, Austria, 1990; Volume 2; Available online: https://inis.iaea.org/records/edxpf-gmx15 (accessed on 27 November 2024).
22. Cereals & Grains Association AACC Approved Methods of Analysis-11th Edition. Available online: https://www.cerealsgrains.org/resources/Methods/Pages/default.aspx (accessed on 16 November 2024).
23. Pena, R.J.; Amaya, A.; Rajaram, S.; Mujeeb-Kazi, A. Variation in Quality Characteristics Associated with Some Spring 1B/1R Translocation Wheats. *J. Cereal Sci.* **1990**, *12*, 105–112. [CrossRef]
24. Vega-Celedón, P.; Canchignia Martínez, H.; González, M.; Seeger, M. Review Biosynthesis of Indole-3-Acetic Acid and Plant Growth Promoting by Bacteria. *Cultiv. Trop.* **2016**, *37*, 33–39. [CrossRef]
25. Timofeeva, A.; Galyamova, M.; Sedykh, S. Prospects for Using Phosphate-Solubilizing Microorganisms as Natural Fertilizers in Agriculture. *Plants* **2022**, *11*, 2119. [CrossRef]
26. Alori, E.T.; Glick, B.R.; Babalola, O.O. Microbial Phosphorus Solubilization and Its Potential for Use in Sustainable Agriculture. *Front. Microbiol.* **2017**, *8*, 971. [CrossRef]
27. Moreno, I.; Ramírez, A.; Plana, R.; Iglesias, L. El Cultivo Del Trigo. Algunos Resultados de Su Producción En Cuba. *Cultiv. Trop.* **2001**, *22*, 55–67.
28. Oksel, C.; Balkan, A.; Bilgin, O.; Mirik, M.; Baser, I. Investigation of the effect of PGPR on yield and some yield components in winter wheat (*Triticum aestivum* L.). *Turk. J. Field Crops* **2022**, *27*, 127–133. [CrossRef]
29. Gaspareto, R.N.; Jalal, A.; Ito, W.C.N.; Oliveira, C.E.d.S.; Garcia, C.M.d.P.; Boleta, E.H.M.; Rosa, P.A.L.; Galindo, F.S.; Buzetti, S.; Ghaley, B.B.; et al. Inoculation with Plant Growth-Promoting Bacteria and Nitrogen Doses Improves Wheat Productivity and Nitrogen Use Efficiency. *Microorganisms* **2023**, *11*, 1046. [CrossRef] [PubMed]
30. Pardo-Díaz, S.; Romero-Perdomo, F.; Mendoza-Labrador, J.; Delgadillo-Duran, D.; Castro-Rincon, E.; Silva, A.M.M.; Rojas-Tapias, D.F.; Cardoso, E.J.B.N.; Estrada-Bonilla, G.A. Endophytic PGPB Improves Plant Growth and Quality, and Modulates the Bacterial Community of an Intercropping System. *Front. Sustain. Food Syst.* **2021**, *5*, 1–10. [CrossRef]
31. Saia, S.; Rappa, V.; Ruisi, P.; Abenavoli, M.R.; Sunseri, F.; Giambalvo, D.; Frenda, A.S.; Martinelli, F. Soil Inoculation with Symbiotic Microorganisms Promotes Plant Growth and Nutrient Transporter Genes Expression in Durum Wheat. *Front. Plant Sci.* **2015**, *6*, 815. [CrossRef]
32. Yokamo, S.; Irfan, M.; Huan, W.; Wang, B.; Wang, Y.; Ishfaq, M.; Lu, D.; Chen, X.; Cai, Q.; Wang, H. Global Evaluation of Key Factors Influencing Nitrogen Fertilization Efficiency in Wheat: A Recent Meta-Analysis (2000–2022). *Front. Plant Sci.* **2023**, *14*, 1–17. [CrossRef]
33. García-Galindo, L.A.; Capera-Rivas, A.; Mayorquín, N.; Melendez, J.P. Alternativas Microbiológicas Para La Remediación de Suelos y Aguas Contaminados Con Fertilizantes Nitrogenados Microbiological Alternatives for the Remediation of Soils and Water Contaminated with Nitrogen Fertilizers. *Sci. Tech. Año XXV* **2020**, *25*, 172–182. [CrossRef]
34. Ren, N.; Wang, Y.; Ye, Y.; Zhao, Y.; Huang, Y.; Fu, W.; Chu, X. Effects of Continuous Nitrogen Fertilizer Application on the Diversity and Composition of Rhizosphere Soil Bacteria. *Front. Microbiol.* **2020**, *11*, 1948. [CrossRef]
35. Guo, W.; Qi, X.; Xiao, Y.; Li, P.; Andersen, M.N.; Zhang, Y.; Zhao, Z. Effects of Reclaimedwater Irrigation on Microbial Diversity and Composition of Soil with Reducing Nitrogen Fertilization. *Water* **2018**, *10*, 365. [CrossRef]
36. Chen, S.; Waghmode, T.R.; Sun, R.; Kuramae, E.E.; Hu, C.; Liu, B. Root-Associated Microbiomes of Wheat under the Combined Effect of Plant Development and Nitrogen Fertilization. *Microbiome* **2019**, *7*, 136. [CrossRef]
37. Ammann, A.B.; Kölle, L.; Brandl, H. Detection of Bacterial Endospores in Soil by Terbium Fluorescence. *Int. J. Microbiol.* **2011**, *2011*, 435281. [CrossRef]
38. Pérez, A.; Bustamante, P.; Rodríguez, C.; Viñals, R. Influencia de La Fertilización Nitrogenada Sobre La Microflora Edáfica y Algunos Indicadores Del Crecimiento y El Rendimiento de Coffea Canephora Pierre Cultivado En Suelo Pardo Ócrico Sin Carbonatos. *Inst. Nac. Cienc. Agrícolas-Cuba* **2005**, *26*, 65–71.

39. Lerner, S.E.; Arata, A.F.; Arrigoni, A.C. Relationship between Nitrogen Use Efficiency and Industrial Quality in Argentinean Bread Wheat Varieties (*Triticum aestivum* L.) with Different Composition of Gluten. *Rev. Investig. Agropecu.* **2016**, *42*, 29–42.
40. Karagöz, H. Determination of the Yield Quality and Winter Durability Characteristics of Some Bread Wheat (*Triticum aestivum*) Genotypes in Pasinler and Erzincan Locations. *Alinteri J. Agric. Sci.* **2020**, *35*, 30–36. [CrossRef]
41. Tabbita, F.; Ortiz-Monasterio, I.; Piñera-Chavez, F.J.; Ibba, M.I.; Guzmán, C. On-Farm Assessment of Yield and Quality Traits in Durum Wheat. *J. Sci. Food Agric.* **2023**, *103*, 5108–5115. [CrossRef]
42. Guzmán, C.; Autrique, J.E.; Mondal, S.; Singh, R.P.; Govindan, V.; Morales-Dorantes, A.; Posadas-Romano, G.; Crossa, J.; Ammar, K.; Peña, R.J. Response to Drought and Heat Stress on Wheat Quality, with Special Emphasis on Bread-Making Quality, in Durum Wheat. *Field Crops Res.* **2016**, *186*, 157–165. [CrossRef]
43. Wan, Y.; Shewry, P.R.; Hawkesford, M.J. A Novel Family of γ-Gliadin Genes Are Highly Regulated by Nitrogen Supply in Developing Wheat Grain. *J. Exp. Bot.* **2013**, *64*, 161–168. [CrossRef]
44. Horváth, C. Storage Proteins in Wheat (*Triticum aestivum* L.) and the Ecological Impacts Affecting Their Quality and Quantity, with a Focus on Nitrogen Supply. *Columella J. Agric. Environ. Sci.* **2014**, *1*, 57–76. [CrossRef]
45. Da Silveira, A.P.D.; Sala, V.M.R.; Cardoso, E.J.B.N.; Labanca, E.G.; Cipriano, M.A.P. Nitrogen Metabolism and Growth of Wheat Plant Under Diazotrophic Endophytic Bacteria Inoculation. *Appl. Soil. Ecol.* **2016**, *107*, 313–319. [CrossRef]
46. Nadeem, M.Y.; Zhang, J.; Zhou, Y.; Ahmad, S.; Ding, Y.; Li, G. Quantifying the Impact of Reduced Nitrogen Rates on Grain Yield and Nitrogen Use Efficiency in the Wheat and Rice Rotation System of the Yangtze River Region. *Agronomy* **2022**, *12*, 920. [CrossRef]
47. Mantelin, S.; Touraine, B. Plant Growth-Promoting Bacteria and Nitrate Availability: Impacts on Root Development and Nitrate Uptake. *J. Exp. Bot.* **2004**, *55*, 27–34. [CrossRef] [PubMed]
48. Hitz, K.; Clark, A.J.; Van Sanford, D.A. Identifying Nitrogen-Use Efficient Soft Red Winter Wheat Lines in High and Low Nitrogen Environments. *Field Crops Res.* **2017**, *200*, 1–9. [CrossRef]

applied sciences

MDPI

Article

Mechanism of Iron Transport in the *Triticum aestivum* L.–Soil System: Perception from a Pot Experiment

Surong Zhang [1,2], Junquan Yang [1,2,*], Daming Wang [1,2], Jihong Liu [1,2], Jianhua Wang [3], Xiaolong Duan [1,2] and Lingzhi Yang [4]

1 Tianjin Centre, China Geological Survey (North China Center for Geoscience Innovation), Tianjin 300170, China; zhangsurong@126.com (S.Z.); wangdaming@mail.cgs.gov.cn (D.W.); smbx.jh@163.com (J.L.); xiaoduan2012@sina.com (X.D.)
2 Tianjin Key Laboratory of Coast Geological Processes and Environmental Safety, Tianjin 300170, China
3 Aerospace Information Research Institute, Chinese Academy of Sciences, Beijing 100101, China; wjh1979118@163.com
4 Hebei Research Center for Geoanalysis, Baoding 071000, China; cszxzgb@163.com
* Correspondence: dap-yangjunquan@163.com; Tel.: +86-022-2416-9032

Abstract: Iron is one of the necessary trace elements for plant growth and the human body. The 'hidden hunger' phenomenon in the human body caused by an imbalance of iron in soil is increasingly prominent. Addressing this issue and optimizing soil through regulatory measures to improve the absorption and utilization of iron by crops has become an urgent priority in agricultural development. This study carries out pot experiments to observe the growth process of *Triticum aestivum* L. under various soil iron environments. Combined with previous research results, the transport mechanism of iron in the soil—*Triticum aestivum* L. system was systematically explored. The results indicate that during the jointing and maturity stages of *Triticum aestivum* L., iron was preferentially enriched in the underground parts; at the maturity stage, the iron content in various organs of *Triticum aestivum* L. shows a trend of increase followed by a decrease with the soil iron content varying in the following sequence: deficient, moderately deficient, medium, moderately adequate, and adequate. The iron-deficient stress environment causes an increase in the effectiveness of rhizosphere iron, resulting in a higher level of iron in the plant stems, leaves, and seeds. Conversely, when the soil iron content is medium or adequate, the effectiveness of rhizosphere iron decreases, leading to a reduction in the iron content in each part of the plant. A concentration gradient of 7.2 mg/kg in the experimental setup is found to be the most favorable to the enrichment of iron in the shoots of *Triticum aestivum* L. plants. The findings of this experiment provide guidance for the fertilization strategy to mitigate iron deficiency symptoms in plants under similar acidic-alkaline conditions of soil, as well as a systematic mechanism reference and basis for studying the soil-plant-human health relationship.

Keywords: *Triticum aestivum* L.; soil; iron; transport; mechanism

1. Introduction

One of humanity's greatest challenges is how to sustainably feed a large population, especially in China [1,2]. *Triticum aestivum* L. is one of the most important grain crops in the world [3], and increasing its yield is an effective way to solve the problem of food shortage caused by rapid population growth [4,5]. As crop yields continue to improve, the usage of traditional organic fertilizers to increase the availability of major nutrients (N, P, and K) has been increasing. However, effective supplementation of trace nutrients necessary for plant growth (particularly for iron) is neglected in large production systems [6–8]. The utilization of these conventional fertilizers, which usually lack iron, can lead to iron deficiency in the soil-plant system. Such a deficiency not only decreases agricultural yields but also impacts human health through "hidden hunger", an insidious form of micronutrient (Fe) deficiency that can hinder growth, weaken immune function, and lead to anemia [9,10].

Iron deficiency in soil-plant systems is a widespread challenge in agricultural production worldwide and a scientific issue that needs to be further addressed.

Iron is a metal element that is widely distributed in nature, with a high abundance in the Earth's crust. The total iron content in soil typically ranges from 1% to 10%. However, it is not readily available because of its low solubility [11]. According to statistics, potential iron deficiency affects approximately 40% of the world's arable land [12]. Iron is an essential trace mineral element for the growth and development of plants. It serves as an integral component of vital enzymes, including cytochrome, cytochrome oxidase, and catalase [13]. When the soil is deficient in iron, it significantly impacts the growth and development of crops [14], leading to reduced yield and quality [15]. Two methods, leaf spraying iron fertilizer and soil micro-fertilizer, are commonly employed to resolve the symptoms of iron deficiency in *Triticum aestivum* L. in soils. It has been shown that the micro-fertilizer of iron carrier chelates is not affected by the pH value and Ca++ ion concentration of soil medium, which is of great significance for calcareous and alkaline soils [2,16]. However, excessive application of soil micro-fertilizer can result in resource wastage and environmental quality issues such as heavy metal pollution in surface soil and contamination of shallow groundwater, posing health risks [17]. For example, excessive iron contamination is widespread (>0.3 mg/L) [18]. Over 50 mg/L iron will cause toxicity to microorganisms and significantly inhibit microbial activity [19]. The threshold of soil micro-fertilizer usage to ensure the improvement of soil quality without wasting resources remains unknown.

The iron ion has an 18-electron shell, and its high electrostatic field makes it readily form complex ions [20]. To date, the physiological mechanism of iron uptake and utilization by plants has been extensively documented in the literature [21–25]. Romheld was the first to propose the Mechanism I and Mechanism II systems of iron uptake for higher plants under iron deficiency [26]. The typical *Triticum aestivum* L. vegetation studied in this research belongs to mechanism II plants, which are herbaceous monocotyledonous grasses capable of adapting to high-pH soil. In an iron-deficient environment, these plants can induce the production of high-iron transporters (MAs) in the *Triticum aestivum* L. rhizosphere and secrete them to activate insoluble iron in the rhizosphere, forming a stable octahedral Fe^{3+} chelate, i.e., Fe^{3+}-MAs, which is then specifically absorbed by the rhizosphere to adapt to iron deficiency. After the plant root system absorbs iron from the environment, the plant senses its iron nutritional status and regulates the distribution of iron through its own signaling and regulatory system [23]. Previous research has indicated that iron typically exists in plants in the form of chelates. The primary iron-containing chelates in plants include citric acid [27,28], nicotinamide [29] and mugineic acids [30]. The transporter citric acid facilitates the transportation of iron from roots to overground parts. Iron obtained from plant roots can be transported to overground parts in the form of citric acid-iron trivalent chelate through xylem for long distances [31], and this process is achieved through the xylem [32]. Iron is transported from old leaves to new leaves by iron transporters in the phloem. Various types of iron transporters in the phloem, such as iron transporter proteins and nicotinamides, have been identified in multiple studies, specifically binding with ferric ions [33].

In summary, based on the research on the biological mechanisms of soil-plant interaction promoting nutrient efficiency and yield improvement, we observed the growth process of *Triticum aestivum* L. under different iron gradients through pot experimental systems, and further systematically explored the transport mechanism of iron in the soil–*Triticum aestivum* L. system. A measure of 7.2 mg/kg of iron is found to be the most favorable to the enrichment of iron in the shoots of *Triticum aestivum* L. plants. This study not only provides an innovative perspective and approach to reveal the promotion of nutrient efficiency plant-soil systems but also provides a scientific theoretical basis for the study of the soil-plant-human health relationship. In addition, this study provides theoretical and technical support for the development of green intelligent trace element fertilizer to promote the development of green agriculture.

2. Materials and Methods

2.1. Experimental Design

This study selected the *Triticum aestivum* L. variety of "Shixin 828", extensively cultivated in the North China Plain, for outdoor pot experiments using its seeds as the culture and analyte subjects. Before planting, the seeds were screened to remove poor-quality grains, ensuring that the seeds were essentially plump. The soil used was natural farmland soil, and the mineral components and contents of the soil samples were as follows: 66% quartz, 11% plagioclase, 6% mica, 5% calcite, 4% potassium feldspar, 3% pyroxene, 2% chlorite, 2% amphibole, and a small amount of zeolite. The basic physical and chemical properties of the soil samples were as follows: pH 8.4, organic matter 13.9 g/kg, nitrogen 830 mg/kg, fast-acting potassium 118 mg/kg and available phosphorus 11.2 mg/kg. After natural air-drying, the soil samples were cleared of gravel and plant residues, sieved through a 10 mm nylon screen, and thoroughly mixed, and then 15 kg of the original soil sample was placed into each pot of the same size. The experimental planting pots measure 35 cm in height and 32 cm in diameter, and each pot contains 15 kg of original soil samples. Following the method of controlled variable experiments, ferrous sulfate ($FeSO_4 \cdot 7H_2O$) was used as the iron source to set six concentration gradient groups of iron (Figures 1 and 2), including an original soil control group. Furthermore, six parallel samples with identical soil and planting conditions were created for each gradient, resulting in a total of 36 experimental soil pots. Throughout the experiment, the planting pattern, sunlight exposure, fertilizer and water application amounts, as well as temperature and humidity within the experimental pots, were maintained under identical conditions, effectively removing external environmental interference within the controlled range.

Figure 1. Layout of iron concentration gradient configuration.

Figure 2. Photos showing experimental plants in the pots.

The concentration gradient configuration of the experimental soil is determined based on the effective iron content in the original soil. The third-order lower limit value of effective iron, as outlined in the Land Quality Geochemical Evaluation Specification (DZ/T 0295-2016) [34], is used as the baseline for the deficient status (Table 1), ensuring that the actual iron deficiency status in the soil can be implemented. In alkaline soil, the concentration of soluble iron in the soil solution decreases as the pH values increase. Research indicates that at higher pH levels, each unit increase in pH results in a 1000-fold reduction in active iron within the solution [35]. The extraction process in the laboratory is adequate, closely approximating the total adsorption capacity of active elements in the soil. However, the actual release amount in alkaline soil generally does not reach the extraction effect of the measured effective status. Hence, given the alkaline nature of the soil in this experiment (pH = 8.4), the available iron content in the experiment is estimated to be reduced by 5 times, resulting in levels generally lower than the critical value of deficiency lower limit. The iron concentration gradient in the experimental soil is determined based on the original third class of the effective index of soil trace elements (DZ/T 0295-2016). The average value is interpolated between the original first, second, and third classes, and then the original first class is doubled, resulting in the establishment of a new fifth-class index (Table 2). Based on the estimated available iron content in the soil, exogenously added ferrous sulfate compensation solutions were prepared at effective iron concentration levels of 0 mg/kg, 4.5 mg/kg, 7.2 mg/kg, 10 mg/kg, 15 mg/kg, and 40 mg/kg, of which 0 mg/kg serves as the control group (CK). For each concentration gradient, six samples were treated with these solutions. The ferrous sulfate compensation solution contains Fe^{2+} as reduced iron ions, which are easily oxidized to Fe^{3+} in the soil solution, and Fe^{3+} tends to form ferric hydroxide precipitates, limiting the availability of iron ions for plant uptake. In this experiment, the compensation solution is chelated with citric acid and the pH is adjusted to neutral with sodium carbonate before being added to the soil. The chelation of citric acid with iron forms a soluble iron complex, facilitating the mobility of iron and allowing iron ions to diffuse to the roots for plant uptake. The soil was uniformly mixed and allowed to stand until it was dry and free from muddy clumps before evenly planting *Triticum aestivum* L. in loosened, leveled soil.

Table 1. Grade index of available iron in soil (based on DZ/T 0295-2016 appendix D1).

Indices	Class A (Adequate)	Class B (Moderately Adequate)	Class C (Medium)	Class D (Moderately Deficient)	Class E (Deficient)
Available iron (mg/kg)	>20	>10~20	>4.5~10	>2.5~4.5	≤2.5

Table 2. Grade index of available iron concentration gradient in the experimental soil.

Indices	Class A (Adequate)	Class B (Moderately Adequate)	Class C (Medium)	Class D (Moderately Deficient)	Class E (Deficient)
Available iron (mg/kg)	>40	>15~40	>10~15	>7.2~10	≤4.5

The sample data collection for this experiment is primarily divided into two stages: the jointing stage and the mature stage of *Triticum aestivum* L. growth. At the jointing stage, five *Triticum aestivum* L. seedlings and their root soil were randomly chosen from each experimental pot and sent to the laboratory for the analyses of the total and effective iron content in the root soils and the iron content in the *Triticum aestivum* L. roots, stems and leaves. At the mature stage, all the *Triticum aestivum* L. seedlings were initially collected from the individual experimental pots. As the characteristics of *Triticum aestivum* L. growth changed, the sampling parts were adjusted to include the root soil, *Triticum aestivum* L. roots, stems, and ears (seeds with hulls). In the process of the experiment, the elevated outdoor temperatures during the grain filling stage of *Triticum aestivum* L. partially hindered the photosynthesis of *Triticum aestivum* L. leaves and affected the synthesis and accumulation of organic matter. Meanwhile, high nighttime temperatures accelerate *Triticum aestivum* L. respiration, resulting in increased nutrient consumption, insufficient grain filling, and reduced yield. To ensure the grain weight of the *Triticum aestivum* L. ear sample, the test sample at the final maturity stage consists of seeds with hulls. Once the samples are collected, the plant samples are promptly separated into different parts (roots, stems, leaves, and seeds), air-dried, and then sent to the laboratory for testing.

2.2. Solution Preparation

First of all, the compensation concentration values that should be applied at different gradient levels were calculated using the following formula:

$$FeC_x = C_{Fex} - C_{Fe0} \quad (1)$$

where FeC_x represents the compensation concentration value that should be applied to the x-th concentration gradient of Fe, C_{Fex} represents the soil standard concentration value corresponding to the x-th concentration gradient of Fe in Table 2, C_{Fe0} represents the available iron content in the soil (i.e., the presumed available iron content in soil for the control group), and $x(x = 1, 2, \ldots, 5)$ represents the corresponding gradient level of the element.

In addition, the concentration of solution to be applied to make up for the compensation concentration (FeC_{sx}) was calculated using the following formula:

$$Fe_M \times FeC_{sx} = S_M \times FeC_x \quad (2)$$

where Fe_M represents the amount of solution applied per pot for each application, calculated as 1.5 kg, FeC_{sx} represents the application concentration required to make up the compensation for the x-th concentration gradient of Fe, and S_M represents the dry weight of the experimental soil, calculated as 15 kg dry weight. From this, the concentration of each micronutrient application level was formulated (Table 3).

Table 3. Calculation of concentration of applied solution (based on 1. 5 kg of liquid per application, unit mg/kg).

Indicator	Measured Available Content in Soil	Presumed Available Content in Soil	Solution Preparation					
Available iron (mg/kg)	11.9	2.38	Target concentration	40	15	10	7.2	4.5
			Compensation concentration	37.62	12.62	7.62	4.82	2.12
			Application concentration	376.2	126.2	76.2	48.2	21.2

Based on the above-calculated concentration of each class of application solution, the amount of ferrous sulfate needed for each class was calculated by applying 1.5 kg of solution per pot, and totaling 9 kg of application solution for each class of 6 pots, taking into account of factors such as spraying losses (Table 4). In order to fully facilitate the absorption of plant roots, citric acid must be used to prepare complex concentrated solution, which will be diluted when applying. The specific operation is as follows: weigh the required weight of each class of ferrous sulfate raw material and put it in a 500 mL beaker (weighed clean and dry beaker in advance, accurate to 0.001 g), add an appropriate amount of water to dissolve, then add citric acid for complexation, and then gradually add sodium carbonate to adjust the pH value to 5–7 after complete dissolution; finally, place the beaker on a balance and make up for the weight of concentrated solution with water.

Table 4. Weights of materials used for each 1 kg of solution preparation (mg).

Element	Raw Materials			Class A	Class B	Class C	Class D	Class E
	Compound	Mol. Formula	At.w./Cont. %	Conc./Weight	Conc./Weight	Conc./Weight	Conc./Weight	Conc./Weight
Available iron	Fe(II) sulfate M.W.	$FeSO_4·7H_2O$ 278.05	55.85 20.09	376.2 1872.57	126.2 628.17	76.2 379.29	48.2 239.92	21.2 105.53

Cont.: content, Conc.: concentration, M.W.: molecular weight.

2.3. Sample Analysis

The sample analysis was conducted by Hebei Geological Testing and Analysis Center. The total iron content in soil samples was determined using an X-ray fluorescence spectrometer (ZSX Primus II, Rigaku Corporation, Tokyo, Japan). The available iron in soil was determined using the DTPA extraction method; extracts were analyzed using an inductively coupled plasma mass spectrometer (ICP-MS, model Agilent 725, Agilent Technologies, Santa Clara, CA, USA). The pH was determined using the glass electrode method, with measurements performed using the PHS-3C pH meter produced by Shanghai Kangyi Instrument Co., Ltd., Shanghai, China. The iron content in plants was determined in accordance with the method requirements outlined in the National Food Safety Standard—Determination of Multiple Elements in Foods (GB5009.268-2016) [36]. The analysis was performed using an inductively coupled plasma mass spectrometer (ICP-MS, model Agilent 7900, Agilent Technologies, Santa Clara, CA, USA). The complete elemental analysis utilizes first-class national reference materials for quality control, ensuring that the analytical data reportable rate, accuracy, and precision qualification rate all achieve 100%.

2.4. Evaluation Indicators

2.4.1. Enrichment Coefficient

The enrichment coefficient is the ratio of the concentration of elements in a specific part of a plant to the concentration of the same elements in the soil in which the plant grows. It is an essential index for describing the accumulation trend of chemical substances in organisms and, to some extent, reflects the challenges associated with elements migrating from sediments or soil to plants [37,38].

The coefficient is formulated as follows:

$$\text{enrichment coefficient} = C_{plant}/C_{soil} \tag{3}$$

where C_{plant} represents the concentration of elements in roots, stems or leaves of plants, and C_{soil} represents the concentration of the corresponding elements in soil, measured in mg/kg.

2.4.2. Translocation Factor

The translocation factor is defined as the ratio of the concentration of elements in the overground parts of plants to the corresponding concentration of elements in the underground parts of plants. It serves as a crucial index for describing the translocation of chemical substances within organisms [39], and to a certain extent, reflects the mobility of elements from the plant root to the overground part.

The factor is formulated as:

$$\text{translocation factor} = C_{overground}/C_{underground} \tag{4}$$

where $C_{overground}$ is the sum of concentrations of elements in roots, stems or leaves of plants, and $C_{underground}$ is the concentration of corresponding elements in soil, measured in mg/kg.

2.5. Data Processing

The experimental data were statistically analyzed and organized using Excel 2010. The significance of the differences between means was tested using Tukey's Honest Significant Difference (HSD) method, and statistical significance letters were assigned. Correlation analysis and graphing were completed using SPSS 20.0 software.

3. Results

3.1. Characteristics of Element Contents in Soil and Different Parts of *Triticum aestivum* L.

In this *Triticum aestivum* L. pot experiment, the roots, stems and leaves of *Triticum aestivum L.* at the jointing stage as well as the roots, stems and grains at the mature stage were collected for measurements of the total and available iron contents in the soil and the iron content in the *Triticum aestivum* L. roots, stems, leaves, and grains during various growth periods. The statistical results are listed in Table 5. The results show that the average iron accumulation values in the roots, stems, and leaves of *Triticum aestivum* L. at the jointing stage are 1262, 179, and 175 mg/kg, respectively, suggesting that iron accumulation in the roots is significantly higher compared to the stems and leaves; the average iron accumulation in the roots, stems, and seeds of *Triticum aestivum* L. at the maturity stage is 1661, 484, and 141 mg/kg, respectively, with the order of accumulation being roots > stems > seeds. Iron accumulation in the roots is highest at different growth stages, and the iron content in roots and stems at the maturity stage is higher than that at the jointing stage, while iron accumulation in grains is the lowest. The total and effective iron contents in soil are higher at the jointing stage than those at the maturity stage.

Table 5. Statistical characteristics of iron contents in soil and different parts of *Triticum aestivum* L. in the pot experiment.

Growth Stage	Analyte	Minimum	Maximum	Mean	Standard Deviation
The jointing stage (n = 36)	Total iron in soil	3.82	3.93	3.87	0.03
	Available iron in soil	9.70	30.40	15.13	5.21
	Root	592	3437	1262	551
	Stem	86	872	179	131
	Leaf	112	401	175	61
The maturity stage (n = 36)	Total iron in soil	3.64	3.75	3.70	0.03
	Available iron in soil	10.20	18.60	13.13	2.26
	Root	801	3848	1661	693
	Stem	187	1017	484	215
	Seed	93	216	141	29

Note: The total iron content in the table is expressed in %, while the other units are in mg/kg.

3.2. Correlation Analysis between Total Element Contents in Soil and Specific Contents in Various Parts of *Triticum aestivum* L.

3.2.1. The Jointing Stage

As shown in Table 6, the correlation analysis between the total and effective Fe contents in soil and the Fe contents in roots, stems and leaves of *Triticum aestivum* L. at the jointing stage reveals an extremely significant correlation ($p < 0.01$), with a Pearson correlation coefficient of 0.691. The correlation coefficient values are high between the iron content in *Triticum aestivum* L. leaves and the total iron and effective iron content in soil, with correlation coefficients of 0.504 and 0.626, respectively, indicating a significant correlation. However, the correlation between the iron content in roots and stems and the iron content in soil is poor. The jointing stage is the vigorous growth period of *Triticum aestivum* L., and the reproductive growth gradually intensifies. During the jointing stage of *Triticum aestivum* L., the transport direction of Fe element is mainly from soil to the leaf tip of the growing plant, and the Fe uptake in leaves mainly comes from available Fe in the soil.

Table 6. Correlation analysis between the total and effective iron contents in soil and the iron contents in different parts of *Triticum aestivum* L. at the jointing stage (Pearson correlation coefficient, $n = 36$).

Iron Content	Total Iron	Available Iron	Root	Stem
Total iron	1	0.691 **	−0.110	−0.271
Available iron	0.691 **	1	−0.058	−0.252
Root	−0.110	−0.058	1	0.039
Stem	−0.271	−0.252	0.039	1
Leaf	0.504 **	0.626 **	0.007	−0.086

** A significant correlation at the 0.01 level (two-tailed).

3.2.2. The Maturity Stage

It can be seen from Table 7 that during the maturity stage of *Triticum aestivum* L. growth, the correlation coefficient between available iron in soil and seeds was 0.351, reaching the significant correlation level in terms of the Pearson correlation coefficient ($p < 0.05$). This suggests that the iron element in seeds primarily originates from the soil. The concentration of effective iron in soil directly influences the iron content in seeds. At the same time, the correlation coefficient between *Triticum aestivum* L. roots and stems is 0.638, indicating a significant correlation ($p < 0.01$).

Table 7. Correlation analysis between the total and available iron content in soil and the iron content in different parts of *Triticum aestivum* L. at the maturity stage (Pearson correlation coefficient, $n = 36$).

Iron Content	Total Iron	Available Iron	Root	Stem	Seed
Total iron	1	−0.079	−0.143	−0.033	−0.102
Available iron	−0.079	1	0.141	0.096	−0.351 *
Root	−0.143	0.141	1	0.638 **	−0.153
Stem	−0.033	0.096	0.638 **	1	−0.303
Seed	−0.102	−0.351 *	−0.153	−0.303	1

* A significant correlation at the 0.05 level (two-tailed). ** A significant correlation at the 0.01 level (two-tailed).

3.3. *Variation in Element Contents in Roots, Stems, Leaves and Grains of* Triticum aestivum L.

The variations in iron content in different organs of *Triticum aestivum* L. at different growth stages were obtained from six concentration gradients in the *Triticum aestivum* L. pot experiment, as illustrated in Table 8 and Figure 3. At the jointing stage, the iron content in the leaves did not significantly differ under various soil conditions, ranging from deficient to adequate available iron backgrounds, and it was only significantly higher than the normal value in the adequate background (Figure 3c). The iron content in the stem generally decreased with the increase in the iron ion concentration in the soil (Figure 3b). Meanwhile, the iron content in the roots displayed a pattern of initially decreasing and then increasing, with the lowest mean value occurring when the iron ion concentration gradient in the soil was at a medium level, i.e., 10 mg/kg level (Figure 3a). At the maturity stage, the iron content in roots, stems, and seeds initially increased and then decreased with the rise in iron concentration in the soil (as shown in Figure 3d–f). At a moderately deficient iron ion concentration gradient in the soil (at the 7.2 mg/kg level), the iron content in roots, stems, and seeds reached its peak, with the average iron content in roots recorded at 2206 mg/kg, in stems at 612 mg/kg, and in seeds at 151 mg/kg.

Table 8. Distribution characteristics of iron content in organs of *Triticum aestivum* L. during jointing and maturity stages.

Concentration Gradient (mg/kg)		The Jointing Stage			The Maturity Stage		
		Roots	Stems	Leaves	Roots	Stems	Seeds
CK (n = 6)	Minimum	918	120	118	816	193	113
	Maximum	3437	306	164	1398	735	180
	Mean	1649	227	142	1098	467	155
	Standard Deviation	945	72	17	261	176	23
	Sig.	a	ab	b	c	ab	a
4.5 (n = 6)	Minimum	763	112	139	863	187	114
	Maximum	1932	212	240	1778	724	155
	Mean	1152	172	177	1364	417	133
	Standard Deviation	426	38	35	383	176	16
	Sig.	ab	ab	b	c	ab	a
7.2 (n = 6)	Minimum	676	135	113	926	501	112
	Maximum	1623	872	222	2926	830	188
	Mean	1145	288	162	2206	612	151
	Standard Deviation	388	288	40	767	166	28
	Sig.	ab	a	b	a	a	a
10 (n = 6)	Minimum	592	88	112	871	225	97
	Maximum	1177	189	201	3848	1017	216
	Mean	975	143	156	2142	553	149
	Standard Deviation	206	45	35	1067	342	45
	Sig.	b	ab	b	ab	ab	a
15 (n = 6)	Minimum	802	97	114	801	267	105
	Maximum	1780	196	190	1764	620	181
	Mean	1167	127	143	1444	353	134
	Standard Deviation	330	36	30	370	134	28
	Sig.	ab	b	b	bc	b	a
40 (n = 6)	Minimum	751	86	192	1257	281	93
	Maximum	2429	180	401	2209	848	150
	Mean	1481	116	271	1709	505	127
	Standard Deviation	609	33	83	320	217	24
	Sig.	ab	b	a	abc	ab	a

Note: Statistical significance of means was tested using Tukey's Honest Significant Difference (HSD) test. In the table, "Sig." denotes the significance of the Tukey statistical test ($p < 0.05$). The letter assignment for statistical significance was as follows: the means in each column were arranged from largest to smallest. The highest mean was designated as 'a', and this mean was compared with each subsequent mean. Means not significantly different were also marked as 'a'. Significantly different means were marked as 'b'. The mean marked 'b' was then used as the standard for comparison with larger means, marking means not significantly different as 'b' until a significant difference was found. This process continued with subsequent letters ('c', etc.) until all means in the column were assigned.

The iron content in the roots of *Triticum aestivum* L. at the jointing and mature stages is significantly higher compared to the stems, leaves, and seeds under the treatment of six different iron concentration gradients in the experimental soil. This suggests that the majority of the iron absorbed by *Triticum aestivum* L. roots is accumulated in the roots, with only a small portion being transported to the overground parts. Throughout the *Triticum aestivum* L. growth period, the continuous supply of rhizosphere iron resulted in higher iron content in the roots at the maturity stage than at the jointing stage, and except the control group, the variation trend of the rhizosphere iron is the most significant under iron deficiency stress conditions.

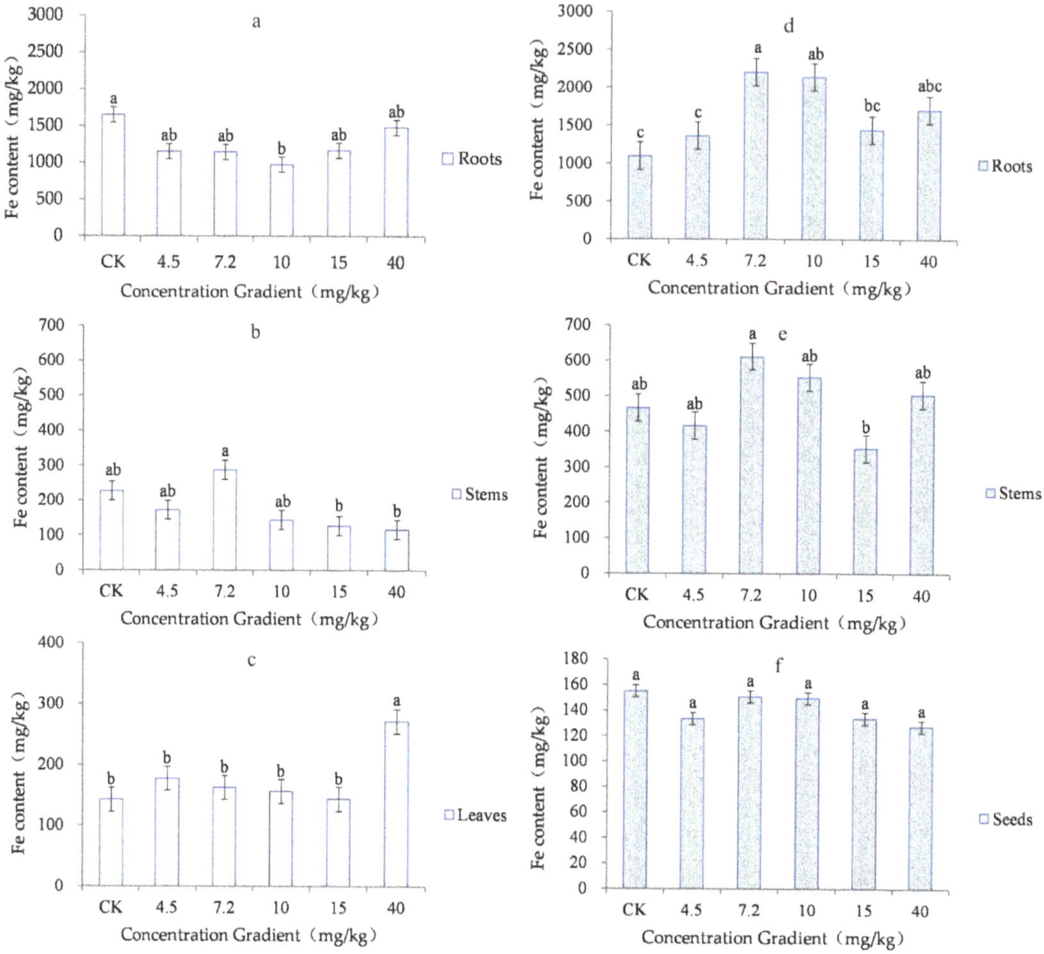

Figure 3. Mean Fe content and its statistical significance in the roots, stems, leaves, and seeds of *Triticum aestivum* L. at the jointing and maturity stages, with significance assessed using the Tukey test. Left: the jointing stage; right: the maturity stage. (**a**) Roots; (**b**) stems; (**c**) leaves; (**d**) roots; (**e**) stems; (**f**) seeds (lowercase letters a, b and c denote significance ($p < 0.05$)).

3.4. Enrichment Coefficients of Iron in Different Parts of Enrichment Coefficients of Iron in Different Parts of Triticum aestivum L.

The enrichment coefficients of iron in different parts of the *Triticum aestivum* L. plants in this experiment are presented in Tables 9 and 10. It can be seen from the table that the enrichment coefficients of different parts of *Triticum aestivum* L. are all less than 1, showing an enrichment pattern of roots > stems/leaves > seeds. The enrichment coefficient of iron in *Triticum aestivum* L. roots is considerably higher than that in overground parts, with the average enrichment coefficient of roots being 0.033 at the jointing stage and 0.045 at the maturity stage, respectively. The enrichment coefficients of stems, leaves, and seeds at the jointing stage are similar, averaging 0.005. At maturity, the enrichment coefficient of the stem and leaf increased by one order of magnitude, reaching an average value of 0.013. The enrichment coefficient of seeds is low, differing by one order of magnitude from that of roots and stems. Conversely, the enrichment coefficient remains largely consistent across different concentration gradients of soil iron. The enrichment coefficient characteristics mentioned above indicate that adjusting the background iron content in the soil has a minimal impact on *Triticum aestivum* L. seeds.

Table 9. Enrichment coefficient and translocation factor of iron in *Triticum aestivum* L. plants at the jointing stage.

Concentration Gradient	Enrichment Coefficient of Root	Enrichment Coefficient of Stem	Enrichment Coefficient of Leaf	Underground to Aboveground Translocation Factor
CK	0.043	0.006	0.004	0.286
4.5 mg/kg	0.030	0.004	0.005	0.321
7.2 mg/kg	0.030	0.007	0.004	0.393
10 mg/kg	0.025	0.004	0.004	0.330
15 mg/kg	0.030	0.003	0.004	0.248
40 mg/kg	0.038	0.003	0.007	0.317
Mean	0.033	0.005	0.005	0.316

Table 10. Enrichment coefficient and translocation factor of iron in *Triticum aestivum* L. plants at the maturity stage.

Concentration Gradient	Enrichment Coefficient of Root	Enrichment Coefficient of Stem/Leaf	Enrichment Coefficient of Seed	Underground to Aboveground Translocation Factor
CK	0.030	0.013	0.004	0.570
4.5 mg/kg	0.037	0.011	0.004	0.415
7.2 mg/kg	0.060	0.017	0.004	0.389
10 mg/kg	0.058	0.015	0.004	0.352
15 mg/kg	0.039	0.010	0.004	0.356
40 mg/kg	0.046	0.014	0.003	0.381
Mean	0.045	0.013	0.004	0.411

4. Discussion

4.1. Mechanism of Iron Transport in Soil and Triticum aestivum L. Plants

The experimental findings indicate that both the total and effective iron contents in the soil decrease as *Triticum aestivum* L. grows, whereas the iron content in various parts of the plant continues to accumulate. The correlation analysis indicates that the available iron in soil predominantly influences the aboveground portions of the plant, such as leaves and grains, signifying the migration of iron between the soil and the *Triticum aestivum* L. plants.

In this pot experiment, citric acid was used as a chelating agent ligand. Based on previous data [40], it is presumed that the transport mechanism of iron between the soil and the *Triticum aestivum* L. plants involves the formation of a cyclic chelate between citric acid and iron ions through carboxyl (COOH) and hydroxyl (OH) coordination groups and its own carbon chain. The strong coordination ability of iron ions enables them to form coordination with the carboxyl and hydroxyl groups in the citric acid ligand. Upon adhering to the root surface, iron ions undergo reduction and degradation on the root surface as a result of the release of H+ and e- by the root system. The iron element is then absorbed into the cells in the form of Fe^{3+} [41] and subsequently transported from the roots to the leaves and seeds through the xylem during the growth of *Triticum aestivum* L.

4.2. Mechanism of Iron Accumulation in Various Parts of Triticum aestivum L. under Iron Deficiency Stress Conditions

This pot planting experiment, conducted under varying soil iron concentrations, revealed that the element contents in various parts of *Triticum aestivum* L. do not show a linear relationship with the background gradient. Particularly at the maturity stage, the iron content in various *Triticum aestivum* L. organs generally displayed an initial increase followed by a decrease as the soil iron content varied in the sequence of deficient, moderately deficient, medium, moderately adequate, and adequate. The iron content in stems, leaves, and seeds reached its peak at the concentration gradient of 7.2 mg/kg, with the soil containing 4.5 mg/kg and 7.2 mg/kg considered deficient and moderately deficient, respectively.

Compared to other concentration gradients, the iron deficiency stress environment accelerates the secretion of plant high iron transporters (MAs) from roots to the rhizosphere and activates insoluble iron. This releases a large amount of plant high iron transporters—mugineic acids (MAs), thereby increasing the availability of Fe in the rhizosphere and facilitating plants to directly absorb the high Fe transporter complex, i.e., Fe^{3+}-PS [22,42,43]. This is achieved through the specific absorption system located on the root plasma membrane, resulting in an increase in iron content in *Triticum aestivum* L. roots, stems, and seeds. Likewise, in a stress environment, when the soil's iron supply is relatively sufficient, it is easier for plant organs to absorb iron, resulting in the highest iron content in all parts of *Triticum aestivum* L. at a concentration of 7.2 mg/kg. However, when the iron content in the soil is moderate or abundant, the effectiveness of iron in the rhizosphere decreases, thus inhibiting the iron uptake by plants to some extent, leading to a decrease in the iron content in various plant organs.

4.3. *Transport and Enrichment of Iron in Triticum aestivum* L.

The average transport coefficients of iron in the jointing stage and mature stage of *Triticum aestivum* L. were 0.316 and 0.411, respectively, both of which were less than 1. This suggests that the transport capacity of iron in the soil by different parts of the *Triticum aestivum* L. plant is low. As *Triticum aestivum* L. grows, its transport capacity generally increases. Specifically, at a concentration gradient of 7.2 mg/kg, the translocation factor from underground to aboveground is the highest during the jointing stage, and the enrichment coefficient of roots and stems corresponding to this concentration gradient is the highest at maturity. Combined with the knowledge obtained by the aforementioned studies on the mechanism of iron accumulation in *Triticum aestivum* L. under iron deficiency stress, the authors suggest that maintaining an iron-deficient soil environment is beneficial for the growth of *Triticum aestivum* L. Previous studies have indicated that the application of iron fertilizer can alleviate iron deficiency symptoms in plants. However, it has been observed that applying iron fertilizer to plant roots or leaves can rapidly desensitize the plant roots' response to iron deficiency stress [44,45]. Therefore, this experiment holds significant reference value for determining the amount of iron fertilizer to be applied to the roots. Given that this experiment was conducted in a highly alkaline soil background, the concentration level of the added iron ion solution should be determined by studying iron transport in various soil backgrounds.

Chlorosis, resulting from iron imbalance in calcareous soil, is widespread in northwest and north China. It affects trees, fruit trees, pastures, vegetables, and field crops, with fruit trees being especially vulnerable to chlorosis compared to annual crops, resulting in a significant decline in fruit product quality. The application of iron micro-fertilizer is a rapid and effective method to alleviate symptoms of iron deficiency in plants. The type of iron fertilizer and its application mode can significantly impact the utilization efficiency of iron in plants. This study can offer valuable insights into enhancing plant iron nutrition by promoting efficient iron uptake and transport through regulatory measures, such as fertilization methods and planting patterns.

5. Conclusions

During both the jointing and maturity stages of *Triticum aestivum* L., there is a preferential enrichment of iron in the underground parts. As *Triticum aestivum* L. undergoes growth, iron ions in the soil establish coordination bonds with carboxyl and hydroxyl groups. Fe^{3+} is continuously transported from the roots to the leaves and seeds via the xylem controlled by the process of root reduction, which facilitates the migration of iron between the soil and *Triticum aestivum* L. At maturity, the iron content in various *Triticum aestivum* L. organs demonstrates a general pattern of initial increase followed by a decrease. This corresponds to the variations in soil iron content categorized as deficient, moderately deficient, medium, moderately adequate, and adequate. This phenomenon can be attributed to the accelerated secretion of high-iron transporter from the *Triticum aestivum* L. root rhizosphere and the enhanced availability of iron in the rhizosphere under conditions of iron deficiency stress. Additionally, the adjustment of background iron content in the soil has minimal influence on the iron content of *Triticum aestivum* L. seeds. The conducted experiment reveals that a concentration gradient of 7.2 mg/kg is most favorable for the enrichment of iron in the shoot of *Triticum aestivum* L. plants. These findings provide valuable insights for developing fertilization strategies aimed at alleviating iron deficiency symptoms in plants growing in similar acidic and alkaline soil conditions.

Author Contributions: Conceptualization, S.Z. and J.Y.; methodology, S.Z.; validation, D.W.; formal analysis, J.L.; investigation, S.Z.; resources, J.Y.; data curation, S.Z. and L.Y.; writing—original draft preparation, S.Z. and J.Y; writing—review and editing, D.W. and X.D.; supervision, J.W. and L.Y.; project administration, S.Z.; funding acquisition, D.W. All authors have read and agreed to the published version of the manuscript.

Funding: This study was supported by the National Natural Science Foundation of China (general project approval no. 42272346) and the project of the China Geological Survey (DD20230101).

Institutional Review Board Statement: Not applicable.

Informed Consent Statement: Not applicable.

Data Availability Statement: Data is contained within the article.

Acknowledgments: The authors would like to express sincere thanks to Wanjun Jiang of Tianjin Centre, China Geological Survey (North China Center for Geoscience Innovation), for his guidance during the paper revision stage, Zhibin Jin of Shanxi Geophysical and Geochemical Exploration Institute Co., Ltd. for his assistance during the solution preparation, Xuesheng Gao and Yunxiao Tong 12 of Tianjin Centre, China Geological Survey, for their efforts in sample collection and promoting the implementation of the experiment, and the project team members for their unwavering support. The authors are grateful to all the experts and editors for their valuable comments on the manuscript and correction of the English text.

Conflicts of Interest: The authors declare no conflicts of interest.

References

1. Li, C.J.; Hoffland, E.; Kuyper, T.; Yu, Y.; Zhang, C.C.; Li, H.G.; Zhang, F.S.; Werf, W. Syndromes of production in intercropping impact yield gains. *Nat. Plants* **2020**, *6*, 653–660. [CrossRef] [PubMed]
2. Wang, N.Q.; Wang, T.Q.; Chen, Y.; Wang, M.; Lu, Q.F.; Wang, K.G.; Dou, Z.C.; Chi, Z.G.; Qiu, W.; Dai, J.; et al. Microbiome convergence enables siderophore-secreting-rhizobacteria to improve iron nutrition and yield of peanut intercropped with maize. *Nat. Commun.* **2024**, *15*, 839. [CrossRef] [PubMed]
3. Wang, Z.; Wang, L.J.; Li, M.X. Study on soil water and salt movement and its effect on wheat production under saline water irrigation in North China Plain. *N. China Geol.* **2023**, *46*, 76–81. (In Chinese) [CrossRef]
4. Ray, D.; Mueller, N.; West, P.; Foley, J. Yield Trends Are Insufficient to Double Global Crop Production by 2050. *PLoS ONE* **2013**, *8*, e66428. [CrossRef] [PubMed]
5. Donaire, G.; Vanzetti, L.; Conde, M.; Bainotti, C.; Mir, L.; Borrás, L.; Chicaiza, O.; Helguera, M. Dissecting genetic loci of yield, yield components, and protein content in bread wheat nested association mapping population. *Euphytica* **2023**, *219*, 65. [CrossRef]
6. Wang, C.Y.; Zhang, S.R.; Liu, J.H.; Xing, Y.; Yang, J.Q. Evaluation of the characteristic land resources with Zn, Se and their ecological effects in Raoyang county of Hebei province. *Geol. Surv. Res.* **2019**, *42*, 49–56. (In Chinese)
7. Fei, X.F.; Lou, Z.H.; Xiao, R.; Ren, Z.Q.; Lv, X.N. Estimating the spatial distribution of soil available trace elements by combining auxiliary soil property data through the Bayesian maximum entropy technique. *Stoch. Environ. Res. Risk Assess.* **2021**, *36*, 2015–2026. [CrossRef]
8. Sindireva, A. Local Biogeochemical Cycles of Trace Elements in Agroecosystems of Western Siberia. *Geochem. Int.* **2023**, *61*, 1048–1060. [CrossRef]
9. Ofori, K.; Antoniello, S.; English, M.; Aryee, A. Improving nutrition through biofortification—A systematic review. *Front. Nutr.* **2022**, *9*, 1043655. [CrossRef]

10. Zulfiqar, U.; Khokhar, A.; Maqsood, M.; Shahbaz, M.; Naz, N.; Sara, M.; Maqsood, S.; Sahar, S.; Hussain, S.; Ahmad, M. Genetic biofortification: Advancing crop nutrition to tackle hidden hunger. *Funct. Integr. Genom.* **2024**, *24*, 34. [CrossRef]
11. Cohen, C.; Fox, T.; Garvin, D.; Kochia, L. The Role of Iron-Deficiency Stress Responses in Stimulating Heavy-Metal Transport in Plants. *Plant Physiol.* **1998**, *116*, 1063–1072. [CrossRef] [PubMed]
12. Zhang, J.; Wu, L.H.; Wang, M.Y. Iron and zinc biofortification in polished rice and accumulation in rice plant (*Oryza sativa* L.) as affected by nitrogen fertilization. *Acta Agric. Scand. Sect. B-Soil Plant Sci.* **2008**, *58*, 267–272. [CrossRef]
13. Ma, F.L.; Song, L.M.; Wang, J.M. Overview of Research on Trace Elements in Soil. *Qinghai Sci. Technol.* **2009**, *16*, 32–36. (In Chinese)
14. Zhang, S.R.; Wang, D.M.; Yang, J.Q.; Zhang, J.; Wang, J.H.; Zhang, D.H.; Tong, Y.X.; Jin, Z.B.; Chen, D.L. Progress and Prospects for Remote Sensing Quantitative Inversion Research on Soil Elements. *Geol. China*. 2023. *accepted* (In Chinese). Available online: https://link.cnki.net/urlid/11.1167.P.20231205.1103.002 (accessed on 30 May 2024).
15. Velemis, D.; Almaliotis, D.; Bladenopoulou, S.; Karapetsas, N. Leaf nutrient levels of apple orchards (cv. Starkrimson) in relation to crop yield. *Adv. Hort. Sc.* **1999**, *13*, 147–150.
16. Dai, J.; Qiu, W.; Wang, N.Q.; Wang, T.Q.; Nakanishi, H.; Zuo, Y.M. From Leguminosae/Gramineae Intercropping Systems to See Benefits of Intercropping on Iron Nutrition. *Front. Plant Sci.* **2019**, *10*, 605. [CrossRef]
17. Jiang, W.J.; Meng, L.S.; Liu, F.T.; Sheng, Y.Z.; Chen, S.M.; Yang, J.L.; Mao, H.R.; Zhang, J.; Zhang, Z.; Ning, H. Distribution, source investigation, and risk assessment of topsoil heavy metals in areas with intensive anthropogenic activities using the positive matrix factorization (PMF) model coupled with self-organizing map (SOM). *Environ. Geochem. Health* **2023**, *45*, 6353–6370. [CrossRef] [PubMed]
18. Liu, Y.; Feng, C.P.; Sheng, Y.Z.; Dong, S.S.; Chen, N.; Hao, C. Effect of Fe(II) on reactivity of heterotrophic denitrifiers in the remediation of nitrate- and Fe(II)-contaminated groundwater. *Ecotoxicol. Environ. Saf.* **2018**, *166*, 437–445. [CrossRef]
19. Liu, Y.; Sheng, Y.Z.; Feng, C.P.; Chen, N.; Liu, T. Distinct functional microbial communities mediating the heterotrophic denitrification in response to the excessive Fe(II) stress in groundwater under wheat-rice stone and rock phosphate amendments. *Environ. Res.* **2020**, *185*, 109391. [CrossRef]
20. Liu, Y.J.; Cao, L.M.; Li, Z.L.; Wang, H.N.; Chu, T.Q.; Zhang, J.R. *Element Geochemistry*, 1st ed.; Science Press: Beijing, China, 1984; pp. 86–88.
21. Marschner, H.; Romheld, V.; Kissel, M. Different strategies in higher plants in mobilization and uptake of iron. *J. Plant Nutr.* **1986**, *9*, 695–713. [CrossRef]
22. Li, L.M.; Wu, L.H.; Ma, G.R. The Progress on Iron-absorbing Mechanism and Related Gene in Plant. *Chin. J. Soil Sci.* **2010**, *41*, 994–999. (In Chinese) [CrossRef]
23. Shen, H.Y.; Xiong, H.C.; Guo, X.T.; Zuo, Y.M. Progress of molecular and physiological mechanism of iron uptake and translocation in plants. *Plant Nutr. Fertil. Sci.* **2011**, *17*, 1522–1530. (In Chinese)
24. Briat, J.; Dubos, C.; Gaymard, F. Iron nutrition, biomass production, and plant product quality. *Trends Plant Sci.* **2015**, *20*, 33–40. [CrossRef]
25. Chang, J.B.; Ma, Z.Y.; Ding, Z.J.; Zheng, S.J. Research progresses on molecular mechanisms of storage, transportation and reutilization of plant seed iron. *J. Zhejiang Univ. (Agric. Life Sci.)* **2021**, *47*, 473–480. (In Chinese) [CrossRef]
26. Römheld, V.; Marschner, H. Mechanism of Iron Uptake by Peanut Plants. *Plant Physiol.* **1983**, *71*, 949–954. [CrossRef]
27. Brown, J.; Chaney, R. Effect of Iron on the Transport of Citrate into the Xylem of Soybeans and Tomatoes. *Plant Physiol.* **1971**, *47*, 836–840. [CrossRef] [PubMed]
28. Tiffin, L.O. Iron Translocation II. Citrate/Iron Ratios in Plant Stem Exudates. *Plant Physiol.* **1966**, *41*, 515–518. [CrossRef]
29. Hell, R.; Stephan, U.W. Iron uptake, trafficking and homeostasis in plants. *Planta* **2003**, *216*, 541–551. [CrossRef] [PubMed]
30. Aoyama, T.; Kobayashi, T.; Takahashi, M.; Nagasaka, S.; Usuda, K.; Kakei, Y.; Ishimaru, Y.; Nakanishi, H.; Mori, S.; Nishizawa, N. OsYSL18 is a rice iron(III)-deoxymugineic acid transporter specifically expressed in reproductive organs and phloem of lamina joints. *Plant Mol. Biol.* **2009**, *70*, 681–692. [CrossRef] [PubMed]
31. Morrissey, J.; Baxter, I.; Lee, J.; Li, L.; Lahner, B.; Grotz, N.; Kaplan, J.; Guerinot, S. The Ferroportin Metal Efflux Proteins Function in Iron and Cobalt Homeostasis in Arabidopsis. *Plant Cell* **2009**, *21*, 3326–3338. [CrossRef]
32. Kim, S.; Guerinot, L.M. Mining iron: Iron uptake and transport in plants. *Febs Lett.* **2007**, *581*, 2273–2280. [CrossRef]
33. Krüger, C.; Berkowitz, O.; Stephan, U.W.; Hell, R. A metal-binding member of the late embryogenesis abundant protein family transports iron in the phloem of *Ricinus communis* L. *J. Biol. Chem.* **2002**, *277*, 25062–25069. [CrossRef] [PubMed]
34. *DZ/T 0295-2016*; Specification for Geochemical Evaluation of Land Quality. Ministry of Land and Resources of the People's Republic of China. Geological Press: Beijing, China, 2016.
35. Sun, X.; Guo, P.C.; Tao, L.N.; Wang, F.W.; Zhang, Y.D.; Li, X.B. *Plant Nutrition and Fertilizers*, 1st ed.; Agricultural Press: Beijing, China, 1988; pp. 154–155.
36. *GB5009.268-2016*; National Food Safety Standards—Determination of Multiple Elements in Food. The National Health and Family Planning Commission of the People's Republic of China and State Food and Drug Administration: Beijing, China, 2016.
37. Zhang, J.J. Study on Monitoring the Change of Heavy Metal Content in Soil Crop System of Farmland by Hyperspectral Remote Sensing. Master's Thesis, Nanjing University, Nanjing, China, May 2019. (In Chinese).
38. Pan, Y.H.; Wang, H.B.; Gu, Z.P.; Xiong, G.H.; Yi, F. Accumulation and translocation of heavy metals by macrophytes. *Acta Ecol. Sin.* **2010**, *30*, 6430–6441.

39. Yan, L.; Li, L.S.; Ni, X.L.; Li, C.X.; Li, J. Accumulation of Soil Heavy Metals in Five Species of Wetland Plants. *Acta Bot. Boreal. Occident. Sin.* **2016**, *36*, 2078–2085. (In Chinese) [CrossRef]
40. Fan, Y.J.; Wei, L.M.; Liang, Z.; Gu, J.W.; Qiu, S.B.; Wang, T.J. A green synthetic route to fischer-tropsch catalysts by complexing and dissolving iron powder with citric acid. *Renew. Energy Resour.* (In Chinese) **2024**, *42*, 448–454. [CrossRef]
41. Bai, Y.J.; Mu, S.H. Uptake and Transport of Iron in Plants and Relations between Iron and Chlorophyll. *J. Hebei Agric. Univ.* **1994**, *17*, 121–125. (In Chinese)
42. Curie, C.; Cassin, G.; Couch, D.; Divol, F.; Higuchi, K.; Jean, M.; Misson, J.; Schikora, A.; Czernic, P.; Mari, S. Metal movement within the plant: Contribution of nicotianamine and yellow stripe 1-like transporters. *Ann. Bot.* **2009**, *103*, 1–11. [CrossRef]
43. Staiger, D. Chemical Strategies for Iron Acquisition in Plants. *Angew. Chem. Int. Ed.* **2002**, *41*, 2259–2264. [CrossRef]
44. Lopez-millan, A.; Morales, F.; Gogorcena, Y.; Abadía, A.; Abadía, J. Iron resupply-mediated deactivation of Fe-deficiency stress responses in roots of sugar beet. *Aust. J. Plant Physiol.* **2001**, *28*, 171–180. [CrossRef]
45. Enomoto, Y.; Hodoshima, H.; Shimada, H.; Shoji, K.; Yoshihara, T.; Goto, F. Long-distance signals positively regulate the expression of iron uptake genes in tobacco roots. *Planta* **2007**, *227*, 81–89. [CrossRef]

applied sciences

MDPI

Article

Quality of Winter Wheat Flour from Different Sowing and Nitrogen Management Strategies: A Case Study in Northeastern Poland

Krzysztof Lachutta and Krzysztof Józef Jankowski *

Department of Agrotechnology and Agribusiness, Faculty of Agriculture and Forestry, University of Warmia and Mazury in Olsztyn, Oczapowskiego 8, 10-719 Olsztyn, Poland; krzysztof.lachutta@ampol-merol.pl
* Correspondence: krzysztof.jankowski@uwm.edu.pl

Abstract: The study analyzed the effect of nitrogen (N) management and different sowing parameters of winter wheat on the flour quality, rheological properties of flour, and bread quality. Flour was obtained from winter wheat grain produced during a field experiment conducted in 2018–2021. The experiment involved three factors: (i) the sowing date (early (3–6 September), delayed by 14 days, and delayed by 28 days), (ii) sowing density (200, 300, and 400 live grains m^{-2}), and (iii) split application of N fertilizer in spring (40 + 100, 70 + 70, and 100 + 40 kg ha^{-1} in the full tillering stage and the first node stage, respectively). A 28-day delay in sowing increased the total protein content of the flour, water absorption capacity of the flour, dough development time and stability, and degree of softening. When sowing was delayed by 14 or 28 days, the crumb density decreased without affecting the loaf volume. A sowing density of 400 grains m^{-2} had a positive impact on the flour color, dough stability, and loaf volume. The flour color and dough stability were enhanced when N was applied at 100 + 40 kg ha^{-1}, respectively. In turn, the total protein content of flour peaked when it was applied at 40 + 100 kg N ha^{-1}. The quality of flour improved when winter wheat was sown at a density of 400 live grains m^{-2} with a delay of 14 or 28 days and supplied with 100 kg N ha^{-1} in the full tillering stage and 40 kg N ha^{-1} in the first node stage.

Keywords: *Triticum aestivum* L.; sowing date; sowing density; N fertilization; flour; protein and ash content; rheological properties; bread quality

1. Introduction

The global population continues to increase, which generates a higher demand for food, mainly plant-based food [1–4]. The global cereal-equivalent food demand is projected to increase by around 10% in 2030 and around 62% in 2050 due to growing social, economic, and demographic pressures [5]. Wheat is one of the most important cereal crops around the world [3,6,7]. The aim of modern crop breeding practices and wheat production technologies is not only to increase grain yields but also to improve grain quality [8–10]. The technological quality of grain produced for baking purposes is determined based on the chemical composition and physicochemical properties of flour [11–15]. The main quality attributes of wheat flour include its color, total protein content, and crude ash content. Wheat flour with desirable quality parameters is obtained by milling high-quality grain or by blending grains from batches with different technological qualities to achieve an end product with specific properties [16]. Color is a very important quality attribute of wheat flour, in particular in flour intended for the production of bread and pasta, because it affects consumer acceptability and the market value of cereal products [17]. The color of flour is influenced mainly by lutein, xanthophyll, β-carotene, and crude ash minerals (phosphorus, potassium, magnesium, and calcium) [18,19]. The processing suitability of flour is partly determined by its crude ash content [20–23]. The crude ash content could be low in flour for baking light cakes and higher in bread flour [24]. In Poland, flours are classified into types based on their crude ash content (for example, type 450 is light cake flour, type 750 is bread flour, and 2000 is whole-wheat flour) [16]. The color of flour is influenced by the coarseness of the grind and ash content (the lower the ash content, the lighter the flour). White flour is produced mainly from the endosperm, the central part of the kernel, where ash content does not exceed 5 g kg^{-1} dry matter (DM). The bran layer (seed coat) surrounding the endosperm is also partly ground during milling. The crude ash content of the seed coat ranges from 60 to 100 g kg^{-1} DM. Therefore, the finer the grind, the more ash is transferred to flour, resulting in its darker color [16]. Protein content is one of the key factors for classifying wheat grain [25]. Flours intended for various purposes differ in their protein content. Flours with a higher protein content (>125 g kg^{-1} DM) are used in bread production, whereas flours with a lower protein content (<95 g kg^{-1} DM) are best suited for baking cakes and biscuits [26]. The rheological properties of flour determine its suitability for the production of various types of baked goods, and these properties are measured with the use of a farinograph [27,28]. A farinograph supports the dynamic measurements of the consistency of dough made from wheat flour and water, as well as changes in the dough properties during mixing. The following rheological properties are measured with a farinograph: the water absorption capacity of flour, dough development time, dough stability, and degree of softening. The water absorption capacity of flour is defined as the amount of water needed to bring the dough to maximum consistency [29,30]. Flours with a high water absorption capacity are used in the production of light and puffy cakes or breads, and they enhance the end products' taste, increase crumb softness, and delay bread staling [30,31]. In turn, flours with a lower water absorption capacity are used in the production of cakes and biscuits [32]. Elastic dough with a low viscosity absorbs more water during mixing than weak dough [30]. Based on flour's water absorption capacity, changes in the dough consistency during development and the degree of dough softening during mixing are registered by the farinograph [28]. The dough development time is influenced by gluten stability [25,33]. A long dough development time is undesirable in the baking industry, because it increases energy consumption during dough mixing. In turn, a very short dough development time is indicative of flour with low gluten quality [34,35]. Dough stability is defined as the time during which dough retains its shape during proofing. A high dough stability is indicative of a high dough strength and tolerance to mixing [33,34]. The degree of dough softening is also an important technological parameter in the baking industry. A high water absorption capacity and low degree of dough softening testify to the high quality of flour. In turn, a high degree of dough softening is indicative of low-quality flour [36], and it inhibits dough fermentation [37]. The baking quality of wheat flour can be comprehensively assessed based on the crumb density and bread loaf volume in a laboratory baking test [38]. These parameters determine the quality and sensory attributes of bread [38,39]. A dense crumb and low bread loaf volume are undesirable from the consumers' point of view [40].

The technological quality of wheat grain is a complex trait that is determined by genetic factors, agroecological conditions, and agronomic practices, including the level of agricultural inputs [41–44]. Nitrogen fertilization is regarded as the key determinant of the technological quality of wheat grain [13,43–46]. Grain quality is affected not only by the N rate but also by the N application method [47]. In Poland, the optimal rate of N fertilizer in winter wheat for achieving grain yields of 8 Mg ha^{-1} is 160–180 kg ha^{-1}, and 40–50% of that rate should be applied in the full tillering stage (FT), 30–40% in the first node stage (FN), and 20–25% in the "flag leaf just visible, but still rolled" stage [13]. Rapid synthesis of gluten proteins begins around 12 days after wheat flowering and ends 35 days after flowering despite the fact that the substrate (amino acids) is still present. For this reason, the total N rate has to be skillfully split into several applications to promote optimal wheat growth [13]. Nitrogen affects mainly the grain yield when applied in the early stages of wheat development, but it improves the quality of grain and flour when applied in successive growth stages [48–50]. The ash content of flour is determined mainly by weather conditions and cultivar, and it is less influenced by the N rate, N splitting, and date of N application [8,18,51,52]. The color of flour is negatively correlated with the ash content, and it is also weakly influenced by N fertilization [8,18,53–59]. An increase in the N rate increases the protein content of flour [60–62]. In turn, the water absorption capacity of flour, rheological properties of dough (development time, stability, and degree of softening), and bread quality (mainly loaf volume) tend to improve when N is applied in the later growth stages of wheat [18,52,56,60,62–68].

The technological quality of wheat grain is also affected by the sowing date [69] and sowing density [66,70–73]. Global climate change has prompted researchers to redefine the agronomic requirements in wheat production, including the optimal sowing dates [74–78]. However, most studies have examined the effects of delayed sowing mainly in the context of winter wheat yields, whereas the impact of delayed sowing on the quality of grain and flour remains insufficiently investigated [79–83]. According to the limited number of studies on the subject, delayed sowing increases the ash content of wheat grain [84,85]. In late-sown stands, grain ripening occurs at higher temperatures, which promotes crude ash accumulation [84]. The ash content is negatively correlated with the color of flour [18], which suggests that delayed sowing could also affect this attribute [86]. The grain of late-sown wheat (and the resulting flour) had a higher protein content [87,88], probably because wheat plants were exposed to higher temperatures during grain ripening, which promoted protein accumulation [89–92]. There is also limited evidence to indicate that the sowing date is associated with the water absorption capacity of flour, dough stability, and bread loaf volume. Delayed sowing decreases the water absorption capacity of flour and dough stability and increases the bread loaf volume [92,93]. The optimal sowing density in wheat production also needs to be redefined due to global climate change. The sowing density is a product of the sowing date and agricultural inputs in the production technology [13]. In Poland, 250–350 grains m^{-2} should be sown between 15 and 20 September [13]. The influence of N management under different sowing strategies on the quality of wheat flour and bread has not been examined in the literature to date. The present study was undertaken to fill in this knowledge gap.

The objective of this study was to evaluate the effect of the sowing date, sowing density, and split application of N fertilizer in the spring on the flour quality (crude ash content, flour color, and total protein content), rheological properties of flour (water absorption capacity, dough development time, dough stability, and degree of softening), and bread quality (loaf volume and crumb density). The present study can contribute to the development of climate-resilient sowing and N management strategies for the production of high-quality winter wheat flour in northeastern Poland.

2. Materials and Methods

2.1. Field Experiment

Flour was obtained from winter wheat cv. 'Julius' in a field experiment conducted during three growing seasons (2018–2021) at the Agricultural Experiment Station (AES) in Bałcyny (53°35′46.4″ N, 19°51′19.5″ E, northeastern Poland) owned by the University of Warmia and Mazury in Olsztyn. The experiment involved three factors. The first factor was the sowing date: early (6 September 2018, 5 September 2019, and 3 September 2020), delayed by 14 days (17–20 September), and delayed by 28 days (1–4 October). The second factor was the sowing density: 200, 300, and 400 live grains m^{-2}. The third factor was the split application of N fertilizer in the spring in the full tillering stage (FT; BBCH 22–25) and in the first node stage (FN; BBCH 30–31) at 40 + 100, 70 + 70, and 100 + 40 kg ha^{-1} (Pulan, Grupa Azoty SA, Puławy, Poland; ammonium nitrate, 34% N). The third split of N fertilizer (40 kg ha^{-1}) was applied in the "flag leaf just visible, but still rolled" stage (BBCH 37) (Pulan, Grupa Azoty SA, Puławy, Poland; ammonium nitrate, 34% N). In the spring, N was applied on 6–12 March (FT), 7–10 May (FN), and 21–30 May (flag leaf just visible, but still rolled).

The experiment had a split-split-plot design (sowing date was assigned to whole plot treatments, sowing density was assigned to subplot treatments, and the split spring application of N fertilizer was assigned to sub-subplot treatments) with three replications. The plot size was 15 m^2 (10 m by 1.5 m). The preceding crop was winter oilseed rape (*Brassica napus* L.). All field treatments that did not constitute the experimental variables were consistent with the agronomic requirements of winter wheat and good agricultural practices. The experimental conditions (soil type and content of plant-available nutrients) and the production technology of winter wheat were described in detail by Lachutta and Jankowski [94] (Tables S1 and S2).

2.2. Flour Quality

Flour was obtained by grinding wheat grain in a laboratory mill (Brabender, Quadrumat Junior, Duisburg, Germany) according to the procedure described by Lachutta and Jankowski [95]. The crude ash content was determined with an NIR System Infratec^{TM} 1241 grain analyzer (FOSS, Hillerod, Denmark) by measuring near-infrared transmittance in the wavelength range of 570–1050 nm. The flour color was evaluated with the use of a MB-3M whiteness meter (Zakład Badawczy Przemysłu Piekarskiego sp. z o. o., Bydgoszcz, Poland) that measured illuminance, i.e., the density of a luminous flux reflected from the flour surface at a wavelength of 565 nm. The total protein content of flour was determined with an AgriCheck instrument (Bruins Instruments, Puchheim, Bayern, Germany) by measuring near-infrared transmittance in the wavelength range of 730–1100 nm.

2.3. Rheological Properties of Dough and Bread Quality

The water absorption capacity and rheological properties of flour (dough development time, dough stability, and degree of softening) were measured with a Brabender farinograph with head type 50 according to Polish Standard PN-EN ISO 5530–1:2015–01 [96]. The bread quality was assessed in a laboratory baking test according to the method described by Klockiewicz-Kamińska and Brzeziński [97]. The dough was prepared in the Teddy Varimixer (Brøndby, Denmark) by mixing 500 g of flour with a moisture content of 15%, water (the amount of water was determined based on the water absorption capacity of flour at 27–28 °C), yeast (3% relative to the amount of water), and salt (1.5% relative to the amount of flour). Water was added in the amount required to achieve a dough temperature of 30 °C. The dough was fermented at a temperature of 30 °C and a relative humidity of 75–80% for 1 h in a proofing cabinet of the DC 32E electric oven (Sveba-Dahlen, Glimek AB, Fristad, Sweden). Dough portions were placed in baking tins and kept in the proofing cabinet at 35 °C for 20–40 min, i.e., the time required for optimal dough development. Bread was baked in the DC 32E electric oven (Sveba-Dahlen, Glimek AB, Fristad, Sweden) at 230 °C for 35 min. The bread loaf volume and crumb density were determined 24 h after baking. The bread loaf volume was determined by the seed displacement method with the use of millet (*Panicum miliaceum* L.) seeds and a 1200 cm^3 Sa-Wy general-purpose volume scanner (Zakład Badawczy Przemysłu Piekarskiego sp. z o. o., Bydgoszcz, Poland). The amount of millet seeds displaced by the bread sample was equal to its volume. The crumb density was determined with the use of a crumb cutter (Zakład Badawczy Przemysłu Piekarskiego sp. z o. o., Bydgoszcz, Poland). A crumb sample with a volume of 27 cm^3 was cut from the center of a bread loaf, at a distance of minimum 1 cm from the crust. The crumb density was calculated using Equation (1).

$$D_c = \frac{W_c}{V_w} \tag{1}$$

where the following is true:

D_c—crumb density (g cm^{-3});

W_c—crumb weight (g);

V_w—volume of the cylinder (27 cm^3).

The flour quality, rheological properties of dough, and bread quality were assessed by Zakład Badawczy Przemysłu Piekarskiego sp. z o. o. in Bydgoszcz, Poland.

2.4. Weather Conditions

Weather conditions in the growing seasons of winter wheat (2018/2019, 2019/2020, and 2020/2021), with special emphasis on the grain ripening stage (mean daily temperature, precipitation, growing degree days, and the Selyaninov hydrothermal index), were described by Lachutta and Jankowski [94] and Lachutta and Jankowski [95] (Tables S3 and S4). In all growing seasons, the mean daily temperature exceeded the long-term average by 1.6–2.3 °C. In each year of the study, the mean daily temperatures exceeded the long-term averages in June, July, and August (by 2.1–5.3, 0.5–4.2, and 0.1–1.6 °C, respectively). In the first (2018/2019) and second (2019/2020) growing season, the total rainfall approximated the long-term average (595.8 mm) (Table S3). The highest grain yields (10.57–10.90 Mg ha^{-1}) were noted in these growing seasons [94]. In the growing season of 2020/2021, the precipitation exceeded the long-term average by 13% (Table S3), and grain yields were 15–18% lower relative to the remaining years of the study [94].

The quality of winter wheat grain is influenced mainly by weather conditions between the milk stage to the fully ripe stage [13]. During the entire field experiment (2018–2021), the temperature and precipitation during grain ripening (BBCH 73–89) differed across years (Table S4). During grain ripening in 2019, 2020, and 2021, the growing degree days (GDDs) were determined at 251–273, 260–299, and 305–354 °C, respectively. In the 2018/2019 season, delayed sowing increased the GDDs by 22 °C during grain ripening (a particularly high increase of 25–94 °C in the GDDs was observed between the dough stage and the fully ripe stage). In turn, in the 2020/2021 season, delayed sowing decreased the GDDs by 31–49 °C during grain ripening. The precipitation levels during grain ripening were determined at 62.3–79.6 mm (2018/2019), 16.2–33.0 mm (2019/2020), and 75.5–97.9 mm (2020/2021). In the 2018/2019 season, winter wheat plants were exposed to higher precipitation (62.3 vs. 76.3–79.6 mm) during grain ripening. In turn, in the 2019/2020 season, precipitation was lower (33 vs. 19.1–16.2 mm) during grain ripening in late-sown stands. In the third year of the study (2020/2021), late-sown stands were exposed to higher precipitation in the milk stage (BBCH 73–83) but lower precipitation during grain ripening (BBCH 83–89). In general, the first and third growing seasons were characterized by the most favorable values of the Selyaninov hydrothermal index during grain ripening (humid spell). In the second growing season, grain ripening occurred during a dry spell (K = 0.40–0.79) (Table S4).

2.5. Statistical Analysis

The results of the conducted measurements (crude ash content, flour color, total protein content, water absorption capacity, dough development time, dough stability, degree of softening, bread loaf volume, and crumb density) were analyzed using the ANOVA with Statistica software, ver. 13 [98]. Post hoc multiple comparisons were performed with the use of Tukey's HSD test at $p \leq 0.05$. The results of the *F*-test for fixed effects in the ANOVA are presented in Table S5. The relationship between meteorological variables and the studied agronomic parameters was evaluated using the linear regression method. The values of Pearson's correlation coefficient (R) were considered significant at $p \leq 0.01$ and $p \leq 0.05$ (Table S6).

3. Results

3.1. Flour Quality

The flour color was negatively correlated with the crude ash content (Table S6). The crude ash content of flour was positively correlated with weather conditions during grain ripening (Selyaninov index in BBCH 73–89) (Figure 1a). In turn, a negative correlation was noted between the flour color and the Selyaninov hydrothermal index during grain ripening (Figure 1b). A dry spell during grain ripening decreased the crude ash content of flour, which had a positive impact on the flour color. As a result, significantly lighter flour (79.9% of the whiteness standard) with a lower crude ash content (6.3 g kg^{-1} DM) was obtained from winter wheat grain produced in a growing season with the least favorable weather conditions during grain ripening (2019/2020) (Tables 1 and S4).

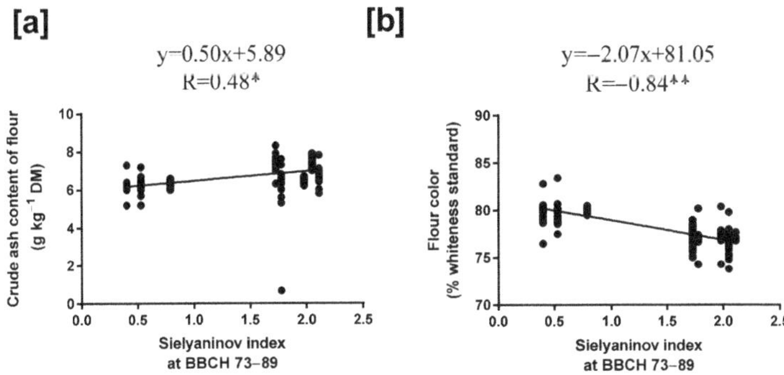

Figure 1. Linear regression between the crude ash content (**a**) and color of flour (**b**) vs. the Selyaninov hydrothermal index during winter wheat grain ripening. * significant at $p \leq 0.05$; ** significant $p \leq 0.01$.

Table 1. Quality parameters of winter wheat flour.

Parameter	Crude Ash Content (g kg^{-1} DM)	Color (% Whiteness Standard)	Total Protein Content (g kg^{-1} DM)	Water Absorption Capacity (%)
		Growing season		
2018/2019	7.4 ^{a}	76.9 ^{b}	135 ^{a}	62.2 ^{a}
2019/2020	6.3 ^{c}	79.9 ^{a}	127 ^{c}	57.7 ^{c}
2020/2021	6.6 ^{b}	77.1 ^{b}	129 ^{b}	59.2 ^{b}
		Sowing date, mean for 2018–2021		
Early	6.8	78.0	129 ^{b}	59.2 ^{b}
Delayed (+14 days)	6.7	77.9	130 ^{b}	59.5 ^{ab}
Delayed (+28 days)	6.7	77.9	132 ^{a}	60.3 ^{a}
		Sowing density (live grains m^{-2}), mean for 2018–2021		
200	6.8	77.9 ^{b}	132 ^{a}	59.6
300	6.8	77.8 ^{b}	130 ^{ab}	59.4
400	6.7	78.2 ^{a}	129 ^{b}	60.0
		Split spring N rate (kg ha^{-1}), mean for 2018–2021		
40 + 100	6.8	77.8 ^{b}	132 ^{a}	59.7
70 + 70	6.7	78.0 ^{ab}	130 ^{ab}	60.0
100 + 40	6.7	78.1 ^{a}	129 ^{b}	59.3

Means with the same letters do not differ significantly at $p \leq 0.05$ in Tukey's test. Means without letters indicate that the main effect is not significant.

The total protein content and water absorption capacity of flour were positively correlated with the crude ash content (Table S6). Therefore, the total protein content and water absorption capacity of flour were higher in years when weather conditions promoted crude ash accumulation (humid and wet spells during grain ripening). The highest total protein content (135 g kg^{-1} DM), the highest crude ash content (7.4 g kg^{-1} DM), and the highest water absorption capacity (62.2%) were noted in flour obtained from grain harvested in the first growing season (2018/2029) (Table 1).

The color of flour was significantly influenced by the sowing density and split spring N rate (Table S5). Lighter flour (78.2%) was obtained from grain grown in plots with the highest sowing density (400 grains m^{-2}) (Table 1). Dense sowing exerted the most beneficial effect on the flour color in the first growing season (Figure 2). A lighter color was also obtained in 70 + 70 and 100 + 40 kg N ha^{-1} (Table 1).

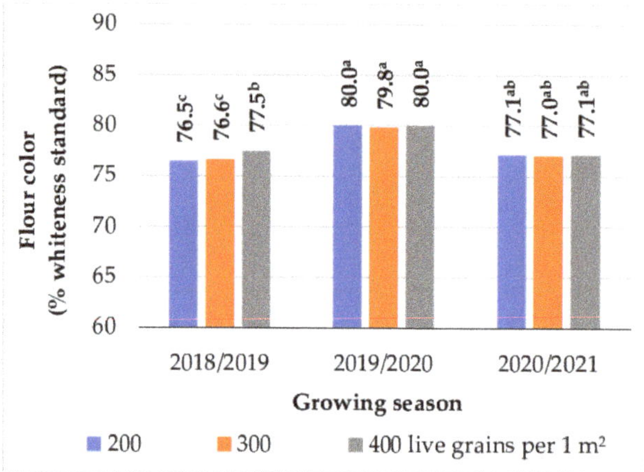

Figure 2. The effect of the sowing density on the color of wheat flour. Means with the same letters do not differ significantly at $p \leq 0.05$ in Tukey's test.

The effect of the sowing date on the total protein content of flour varied depending on weather conditions in the years of the study. Delayed sowing had a particularly positive effect on the total protein content of flour in the first growing season (Figure 3), which was characterized by the most favorable weather conditions during grain ripening (highest values of the Selyaninov index, Table S4). In this year, grain ripening in late-sown stands took place under more supportive weather conditions (humid spell to wet spell, Table S4), which increased the total protein content of flour by 3.8–8.5% (Figure 3). In the remaining growing seasons (year 2 and 3), the sowing date had no effect on the total protein content of flour. An increase in the N rate at FT with a simultaneous decrease in the N rate at FN (100 + 40 kg ha^{-1}) decreased the total protein content of wheat flour by 2.3% (Table 1).

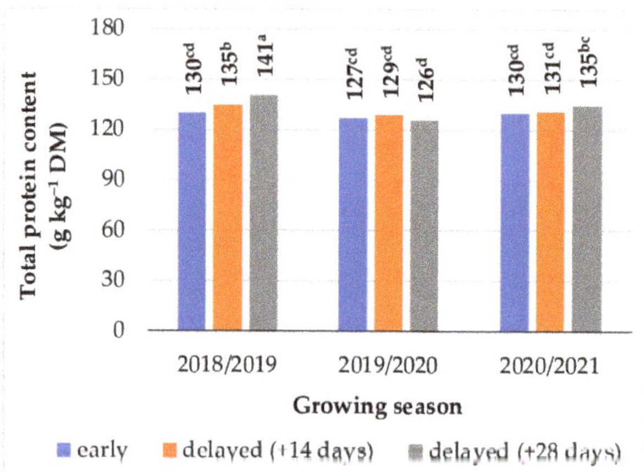

Figure 3. The effect of the sowing date on the total protein content of wheat flour. Means with the same letters do not differ significantly at $p \leq 0.05$ in Tukey's test.

3.2. Rheological Properties of Dough and Bread Quality

The dough stability was negatively correlated with the crude ash content of flour (Table S6) and weather conditions during grain ripening (Figure 4a). In turn, the degree of dough softening was positively correlated with the crude ash content and weather conditions from the milk stage to the fully ripe stage (Figure 4b).

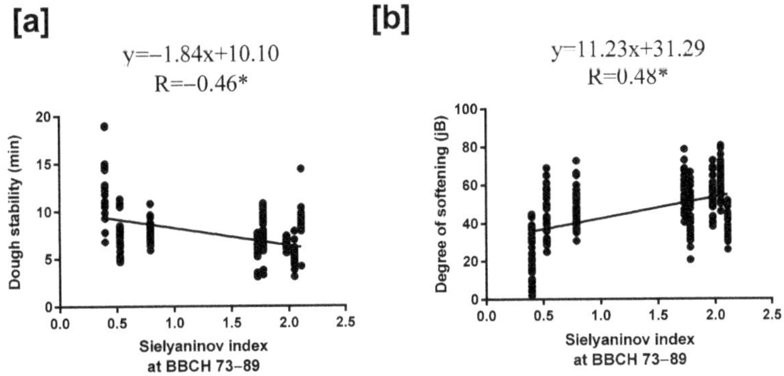

Figure 4. Linear regression between dough stability (**a**) and the degree of softening (**b**) vs. the Selyaninov hydrothermal index during winter wheat grain ripening. * significant at $p \leq 0.05$.

Flour in the second growing season was characterized by more desirable rheological properties, including the longest dough development time (3.7 min), the highest dough stability (9.0 min), and the lowest dough mixing tolerance index (38.4 jB). In this growing season, a dry spell during grain ripening (K = 0.40–0.79) improved the rheological properties of flour (Table S4). The values of dough development time (3.5 min), dough stability (6.3–7.4 min), and degree of softening (48.2–53.5 jB) were least desirable when winter wheat was sown in August (sown early and sown with a delay of 14 days). Delayed sowing (+28 days) had a positive impact on the dough development time and stability (increase of 6% and 14–33%, respectively) and the degree of softening (the dough mixing tolerance index decreased by 12–21%), regardless of weather conditions (Table 2). Delayed sowing improved the rheological properties of flour, because grain ripening took place under less favorable weather conditions (Table S4), which increased the dough stability (negative correlation) and decreased the degree of dough softening during mixing (positive correlation) (Figure 4). The sowing density and split spring N rate exerted a significant effect only on the dough stability (Table S5). The dough stability was highest when winter wheat was sown at a density of 400 grains m^{-2} (7.8 min) and when N fertilizer was applied at 100 and 40 kg ha^{-1} (7.7 min). A decrease in the sowing density to 200–300 grains m^{-2} and the application of 40 + 100 kg N ha^{-1} decreased the dough stability by 8–10% and 8%, respectively (Table 2). An increase in the sowing density to 400 grains m^{-2} had a particularly beneficial influence on the dough stability when sowing was delayed +28 days (Figure 5). The effect of the sowing density and split spring N rate on dough stability did not vary depending on weather conditions in the years of the study (Table S5).

Table 2. Rheological properties of dough and bread quality.

Parameter	Development Time (min)	Stability (min)	Degree of Softening (jB)	Bread Loaf Volume (cm^3)	Crumb Density (g cm^{-3})
		Growing season			
2018/2019	3.5 ^{b}	5.2 ^{c}	60.5 ^{a}	333 ^{b}	0.27 ^{a}
2019/2020	3.7 ^{a}	9.0 ^{a}	38.4 ^{c}	350 ^{a}	0.24 ^{b}
2020/2021	3.5 ^{b}	7.8 ^{b}	45.2 ^{b}	347 ^{a}	0.23 ^{c}
		Sowing date, mean for 2018–2021			
Early	3.5 ^{b}	7.4 ^{b}	48.2 ^{a}	343	0.25 ^{a}
Delayed (+14 days)	3.5 ^{b}	6.3 ^{c}	53.5 ^{a}	343	0.24 ^{b}
Delayed (+28 days)	3.7 ^{a}	8.4 ^{a}	42.4 ^{b}	344	0.24 ^{b}

Table 2. *Cont.*

Parameter	Development Time (min)	Stability (min)	Degree of Softening (jB)	Bread Loaf Volume (cm^3)	Crumb Density (g cm^{-3})
Sowing density (live grains m^{-2}), mean for 2018–2021					
200	3.6	7.2 [b]	49.5	342 [b]	0.25
300	3.5	7.0 [b]	47.9	342 [b]	0.25
400	3.5	7.8 [a]	46.7	347 [a]	0.24
Split spring N rate (kg ha^{-1}), mean for 2018–2021					
40 + 100	3.6	7.1 [b]	48.9	345	0.24
70 + 70	3.5	7.3 [ab]	48.7	343	0.25
100 + 40	3.5	7.7 [a]	46.5	343	0.25

Means with the same letters do not differ significantly at $p \leq 0.05$ in Tukey's test. Means without letters indicate that the main effect is not significant.

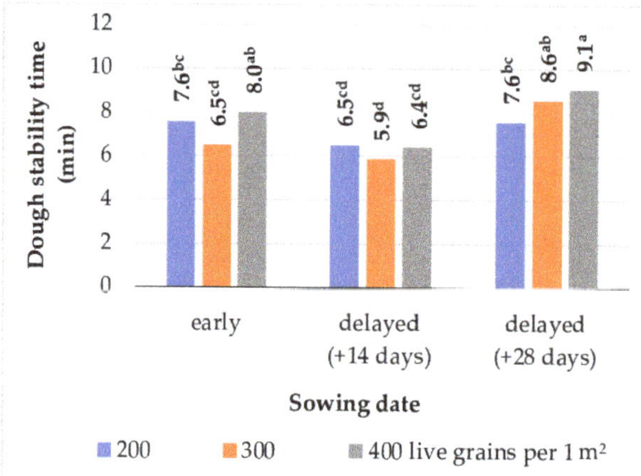

Figure 5. The effect of the sowing date and sowing density on dough stability. Means with the same letters do not differ significantly at $p \leq 0.05$ in Tukey's test.

The analysis of the flour quality also involved a direct assessment of the baking quality in a laboratory baking test. It was assumed that the baking test would reveal specific quality traits that could not be determined with the use of indirect measurement methods. The bread loaf volume and crumb density were influenced by the ash content of the flour. An increase in the ash content decreased the bread loaf volume and increased the crumb density (Table S6). The bread loaf volume was negatively correlated with precipitation in the dough stage (BBCH 83–89). In turn, the crumb density was negatively correlated with the mean daily temperature during grain ripening (BBCH 73–89) (Figure 6). Bread baked from grain harvested in the second and third growing seasons was characterized by the largest loaf volume (347–350 cm^3) and a low crumb density (0.23–0.24 g cm^{-3}). These growing seasons were characterized by the lowest precipitation in the dough stage and the highest mean daily temperatures during grain ripening (Table S4).

A sowing delay of 14 and 28 days decreased the crumb density by 4% (to 0.24 g cm^{-3}) but had no effect on loaf volume (Table 2). The crumb density was not affected by the sowing date only in the second growing season (Figure 7). Delayed sowing decreased the crumb density due to higher mean daily temperatures during grain ripening (Table S4). A high sowing density (400 grains m^{-2}) increased the loaf volume by 1.5% (Table 2). A higher sowing density in the second and third growing seasons induced the greatest improvement in the flour quality (loaf volume increased by 2%) (Figure 8). The loaf volume and crumb density were not affected by the split spring N rate (Table S5).

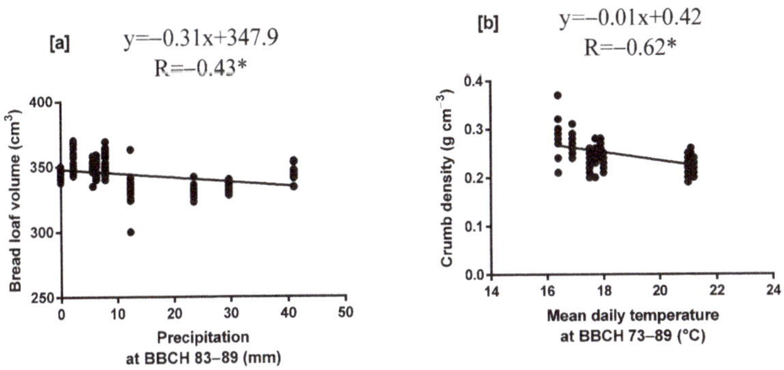

Figure 6. Linear regression between (**a**) bread loaf volume and total precipitation in the dough stage; (**b**) crumb density and mean daily temperature during winter wheat grain ripening. * significant at $p \leq 0.05$.

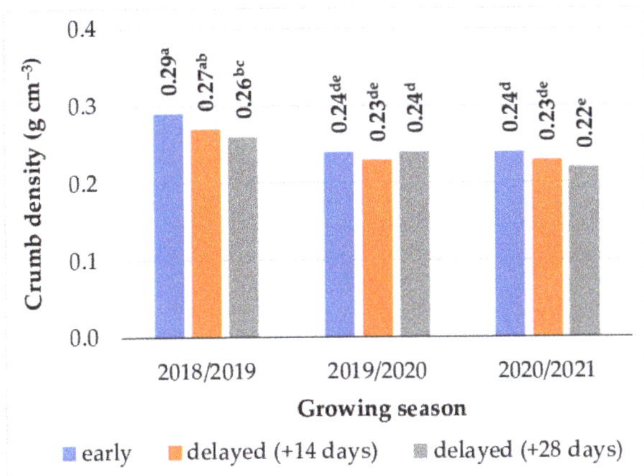

Figure 7. The effect of the sowing date on crumb density. Means with the same letters do not differ significantly at $p \leq 0.05$ in Tukey's test.

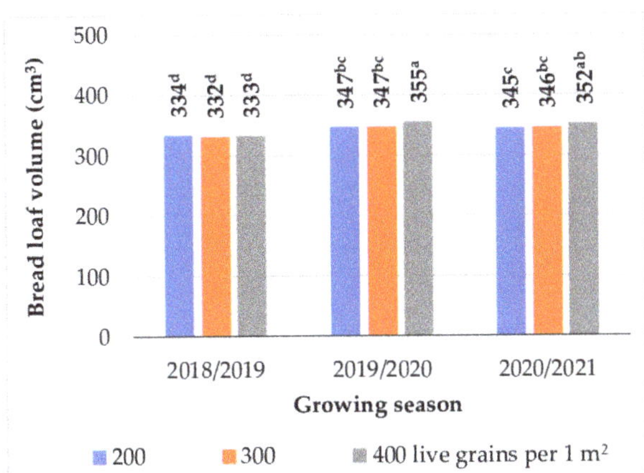

Figure 8. The effect of the sowing density on bread loaf volume. Means with the same letters do not differ significantly at $p \leq 0.05$ in Tukey's test.

4. Discussion

4.1. Flour Quality

The ash content denotes the concentration of minerals in flour [20–23]. The ash content not only affects the nutritional value of flour, but it also determines its technological quality and suitability for the production of various baked goods (for example, ash content does not exceed 5 g kg^{-1} DM in cake flour, ranges from 7.0 to 7.8 g kg^{-1} DM in bread flour, and exceeds 20 g kg^{-1} DM in graham flour) [16]. The ash content of wheat grain is determined mainly by weather conditions, including temperature and precipitation [71,99–101]. This observation was corroborated by the present study, which revealed a positive correlation between the crude ash content of flour and weather conditions during grain ripening. Unfavorable weather conditions during grain ripening decreased the accumulation of crude ash in the flour. The ash content of wheat flour was not influenced by the sowing date, sowing density, or split spring N rate, which is consistent with the findings of other authors [8,18,43,51,52,102–104]. In turn, Adeel et al. [101] and Caglar et al. [71] found that the ash content of flour decreased when sowing was delayed and when the sowing density was increased by 20% and 15%, respectively. According to Caglar et al. [71], Alignan et al. [84], and Adeel et al. [101], delayed sowing decreased the crude ash content of flour because winter wheat was exposed to high temperatures during grain ripening. In the current study, the flour color was very light (77–80%) due a low crude ash content (6.3–7.4 g kg^{-1} DM). Significantly lighter flour was obtained from the grain sown at the highest density of 400 grains m^{-2}. The flour color was also correlated with the sowing density in the work of Caglar et al. [105]. In the present study, the split spring N rate affected the color of flour. Higher N rates applied in FT (100 + 40 kg ha^{-1}) increased flour lightness (78.1%). The application of a portion of the split spring N rate in the FN stage with a simultaneous decrease in the N rate applied in the FT stage (70 + 70 and 40 + 100 kg ha^{-1}) decreased flour lightness by 0.3 percent points (%p). In a study by Jankowski et al. [43], an increase in the N rate combined with systemic fungicide treatment decreased flour lightness by 0.5%p. In turn, the N rate had no effect on the color of wheat flour in the works of Jaskulska et al. [8] and Rodrighero et al. [18].

The protein content of flour determines its baking quality by improving the dough viscosity, extensibility, strength, and elasticity [65,106]. In this study, the total protein content was highest (132 g kg^{-1} DM) in flour obtained from the grain of late-sown winter wheat (+28 days). A similar relationship between the sowing date and protein content of flour was reported by Bagulho et al. [107], Knapowski et al. [92], and Adeel et al. [101]. The observed increase in the protein content of flour can be attributed to the fact that late-sown wheat is exposed to higher temperatures in the ripening stage, which promotes protein accumulation [90,91,108]. In the works of Geleta et al. [109], Mikos-Szymańska and Podolska [110], and Hao et al. [111], and in the present study (Table 1), an increase in sowing density decreased the protein content of flour by 12%, 2%, 6%, and 2%, respectively. In contrast, in a study by Madan and Munjal [112], the protein content of flour was not affected by the sowing density. Xue et al. [65] demonstrated that an increase in the N rate applied at the beginning of stem elongation decreased the protein concentration in flour by 6%. A reverse relationship was noted in this study, where the protein content of flour increased by 2% in response to a higher N rate at the beginning of stem elongation. These differences could be attributed to nutrient dissolution with an increase in grain yields [13]. In the work of Xue et al. [65], an increase in the N rate at the beginning of stem elongation increased the wheat grain yields by 2% and decreased the total protein content of flour. In turn, in our previous study [94], an increase in the N rate applied in the FN stage decreased the grain yield and increased the total protein content of flour. Johansson et al. [113] and Rossmann et al. [114] also found that late N application increased the protein content of wheat grain by 11% and 10%, respectively. In turn, Luo et al. [115], Madan and Munjal [112], and Haile et al. [116] found that the protein content of flour was influenced by the N rate but not by N splitting.

The water absorption capacity of flour is an important parameter that affects the quality of baked goods [117,118]. In the present study, delayed sowing increased the water absorption capacity of flour (by 1.1%p), and similar results were reported by Zhang et al. [119] (increase of 3.4%p). The observed increase in the water absorption capacity of flour could be due to the fact that late-sown wheat plants were exposed to high temperatures during grain ripening, which contributed to protein accumulation at the expense of the synthesis and storage of carbohydrates [99,120]. According to Huang et al. [121], delayed sowing modifies the starch quality by affecting its crystallinity and improving its pasting characteristics, but this effect may be reversed by an excessive delay in sowing [122]. In the works of Knapowski et al. [92] and Caglar et al. [105], delayed sowing decreased the water absorption of flour by 1.1 and 3.9%p, respectively. The sowing density has a weak influence on the water absorption capacity of flour [[123–126] and present study, Table 1], most likely due to the minor effect of this agronomic practice on the protein and starch contents of grain [127–129]. The only study where a higher sowing density increased the water absorption capacity of flour (by 2%) was conducted in northeastern China by Hao et al. [111]. The increase in the water absorption capacity observed by the cited authors [111] could be due to a beneficial influence of a higher seeding rate on the starch content of wheat grain. The higher the starch content of wheat flour, the greater its ability to absorb water during dough mixing [130,131]. Therefore, more water may be required to achieve the optimum dough consistency when using wheat flours with a higher starch content [120]. Split and late N applications can increase the water absorption capacity of flour [64,65,123]. Xue et al. [65] found that the water absorption capacity of flour increased by 3%p when the total N rate was applied in three splits (BBCH 00, BBCH 30, and BBCH 45) rather than two splits (BBCH 00 and BBCH 30). In turn, Blandino et al. [64] reported that foliar N application during flowering (5 kg ha^{-1}) and soil N application at the beginning of heading (40 kg ha^{-1}) increased the water absorption capacity of flour by 2% and 4%, respectively. In the current study and in the work of Warechowska et al. [52], split N application had no effect on the water absorption capacity of flour. In the work of Warechowska et al. [52] and in the present study, the absence of correlations between N splitting and the water absorption capacity of flour could be attributed to the relatively low total protein content of winter wheat grain (128–139 and 129–132 g kg^{-1} DM, respectively) as well as the weak effect of N fertilization on the protein content of grain (±3–5 and 2%, respectively). In a study by Blandino et al. [64], the total protein content of wheat grain was 4–9% higher, and N fertilization induced much greater differences in this parameter (±8%) than in the work of Warechowska et al. [52] and the present study.

4.2. Rheological Properties of Dough and Bread Quality

In the present study, delayed sowing (+28 days) increased the dough development time by 6%. A similar relationship between the sowing date and dough development time was reported by Zhang et al. [119] in China. The influence of sowing density on the dough development time appears to be more ambiguous. Gawęda et al. [126] did not observe changes in the dough development time in response to an increase in the sowing density. In the current study, conducted in northeastern Poland, the sowing density of winter wheat did not induce significant differences in the time of dough development. In turn, Han and Yang [124] and Hao et al. [111] found that a higher sowing density shortened the dough development time. In the work of Zhang et al. [119], the dough development time was influenced by the interaction between the sowing density and N rate. In their study, an increase in the sowing density shortened the dough development time in the absence of N fertilization, but the N rate of 240 kg ha^{-1} prolonged dough development. Jankowski et al. [43] also found that an increase in the spring N rate prolonged dough development by 38%. Blandino et al. [64] demonstrated that late N application had a positive impact on dough development. In their study, a supplemental N rate of 40 kg ha^{-1} applied at the beginning of heading prolonged dough development by 49%, whereas foliar N applied during full flowering prolonged dough development by 32%. In the present study, splitting the N fertilizer rate had no significant influence on the dough development time. According to Jankowski et al. [43] and Blandino et al. [64], N fertilization prolonged the dough development time mainly due to an increase in the total N rate. In the present study, the spring N rate was constant (140 kg ha^{-1}), but N was applied in different splits, which could explain the absence of variations in the dough development time.

In the current experiment, the dough stability was strongly influenced by the sowing date, sowing density, and split spring N rate. The dough stability was highest when winter wheat was sown late (+ 28 days) at the highest density (400 grains m^{-2}) and supplied with 100 + 40 kg N ha^{-1}. In the work of Zhang et al. [119], the dough stability increased by 17% when sowing was delayed by 50 days. In turn, Dong et al. [93] reported that early sowing, a high sowing density, and a high N rate applied in FN had a positive impact on dough stability. Contrary to Zhang et al. [119] and the present study, Hao et al. [111] and Soofizada et al. [73] found that an increase in the sowing density had a negative effect on the dough stability. In the work of Zhang et al. [132], the effect of the sowing density on dough stability was determined by the N rate. The cited authors found that the dough stability increased with a rise in the sowing density only in response to a high N rate (240 kg ha^{-1}). The dough stability was also affected by the sowing density in the experiment performed by Han and Yang [124], where an increase in the sowing density from 90 to 270 grains m^{-2} decreased the dough stability time by 16%. Zhang et al. [119] reported that the dough stability time was prolonged by 42% in response to an N rate of 240 kg ha^{-1} relative to the treatment without N fertilization. Jankowski et al. [43] also demonstrated that the dough stability was nearly three times higher in a high-input production technology of winter wheat. In contrast, Rodrighero [18], Blandino et al. [64], Souza et al. [56], Xue et al. [65], Keres et al. [62], and Cesevičienė et al. [68] did not report any correlations between N fertilization and dough stability.

Zhang et al. [119] found that the degree of dough softening decreased with a delay in sowing, which was also observed in the present study. A relationship between the sowing density and degree of dough softening was not noted in this experiment. The degree of dough softening was not affected by the sowing density in the work of Biel et al. [104] either. Gawęda et al. [123] observed that the degree of dough softening was strongly influenced by the interaction between the sowing density and weather conditions. In their study, a higher sowing density increased the degree of dough softening only in a year characterized by low total precipitation in June. The effect of N fertilization on the degree of dough softening is ambiguous [[62,133,134]; present study, Table 2]. In the work of Keres et al. [62] and in this study, N fertilization did not affect the degree of dough softening. In turn, Kunkulberga et al. [134] reported an increase in the degree of dough softening with an increase in the N rate, whereas Fleitas et al. [133] found that the degree of dough softening decreased in response to higher N rates. In a study by Jankowski et al. [43], the degree of dough softening decreased by 27% in a high-input production technology of winter wheat grain.

The crumb density and bread loaf volume are largely responsible for the taste of bread [38,39]. Consumers have a preference for bread with a large loaf volume and low crumb density [40]. Crumb density is a parameter that describes bread porosity, and it is associated with the properties of gluten [135]. In the present study, bread quality evaluated based on the crumb density and loaf volume was influenced significantly only by the sowing date and sowing density. The crumb density was lower (0.24 g cm^{-3}) when winter wheat was sown in mid-September (+14 days). Delayed sowing decreased the crumb density, because wheat plants were exposed to higher mean daily temperatures during grain ripening. In turn, the loaf volume peaked (347 cm^3) when winter wheat was sown at 400 grains m^{-2}. Knapowski et al. [136] found no relationship between the bread loaf volume and the sowing date of wheat. In contrast, Dong et al. [93] demonstrated that bread made from flour obtained from the grain of early-sown wheat was characterized by the largest loaf volume (808 cm^3). The loaf volume decreased by 11% when sowing was delayed by 21 days. In a study by Zhang et al. [132], the bread loaf volume increased by 7% when the sowing density was increased from 120 to 240 grains m^{-2}. The wheat sowing density had a beneficial influence on the bread loaf volume in the work of Dong et al. [93], whereas no correlation between these parameters was noted by Guerrini et al. [66]. Nitrogen fertilization of wheat increases the bread loaf volume [93,137] but has an undesirable effect on the crumb density (by decreasing porosity) [66]. In the present study, split spring N application had no effect on the bread quality. The study demonstrated that the ash content of flour influences the bread quality. An increase in the ash content decreased the bread loaf volume and increased the crumb density. Nitrogen splitting did not induce differences in the ash content of flour, which could explain why this parameter had no effect on the bread quality.

5. Conclusions

The baking quality of flour was determined mainly by the sowing date, which influenced the highest number of the analyzed quality parameters. The quality attributes of wheat flour were less affected by the sowing density and split spring N rate. Flour of a higher baking quality was obtained from the grain of winter wheat sown between mid-September and early October (with a delay of 14 and 28 days). Delayed sowing increased the total protein content and water absorption capacity of flour and improved the dough development time, dough stability, and degree of softening. Delayed sowing decreased the crumb density (a desirable trait) without compromising the bread loaf volume. The highest sowing density of 400 grains m^{-2} improved the color of flour (without decreasing the crude ash content), dough stability, and bread loaf volume. The total protein content of flour peaked when a portion of the split spring N rate was 40 + 100 kg ha^{-1}. In turn, an increase in the N rate in FT (100 + 40 kg ha^{-1}) had the most beneficial influence on the color of flour and dough stability. The flour quality, rheological properties of dough, and bread quality were determined mainly by the crude ash content of the flour. A low ash content was accompanied by a decrease in the total protein content and water absorption capacity of flour. However, a low crude ash content had a positive effect on the flour color, dough stability, degree of dough softening, bread loaf volume, and crumb density. Therefore, agronomic treatments that decrease the ash content of flour may contribute to improving bread quality by exerting a beneficial influence on the rheological properties of dough. Sowing winter wheat at 400 grains m^{-2} with a delay of 14 or 28 days and the application of 100 + 40 kg N ha^{-1} in the FT stage and in the FN stage, respectively, decreased the crude ash content of flour.

Supplementary Materials: The following supporting information can be downloaded at: https://www.mdpi.com/article/10.3390/app14125167/s1, Table S1. Chemical properties of the analyzed soil; Table S2. Production technology of winter wheat; Table S3. Weather conditions during the growing seasons of winter wheat in 2018–2021 and the long-term average (1981–2015) at the experimental site in the AES in Bałcyny (PM Ecology automatic weather station; PM Ecology Ltd., Gdynia, Poland); Table S4. Phenological development of winter wheat and weather conditions (2018/2019, 2019/2020, and 2020/2021); Table S5. *F*-test statistics in ANOVA; Table S6. Pearson's correlation coefficients denoting the relationships between wheat flour parameters.

Author Contributions: Conceptualization, K.L. and K.J.J.; methodology, K.L. and K.J.J.; software, K.L.; validation, K.L.; formal analysis, K.L.; investigation, K.L.; resources, K.L.; data curation, K.L.; writing—original draft preparation, K.L.; writing—review and editing, K.L. and K.J.J.; visualization, K.L.; supervision, K.J.J.; project administration, K.L. and K.J.J.; funding acquisition, K.J.J. All authors have read and agreed to the published version of the manuscript.

Funding: The results presented in this paper were obtained as part of a comprehensive study financed by the University of Warmia and Mazury in Olsztyn (grant No. 30.610.013–110), funded by the Minister of Science under "the Regional Initiative of Excellence Program".

Institutional Review Board Statement: Not applicable.

Informed Consent Statement: Not applicable.

Data Availability Statement: The original contributions presented in the study are included in the article, further inquiries can be directed to the corresponding author.

Acknowledgments: We would like to thank the staff of the AES in Bałcyny for their technical support during the experiment.

Conflicts of Interest: The authors declare no conflicts of interest.

References

1. Godfray, H.C.J.; Beddington, J.R.; Crute, I.R.; Haddad, L.; Lawrence, D.; Muir, J.F.; Pretty, J.; Robinson, S.; Thomas, S.M.; Toulmin, C. Food security: The challenge of feeding 9 billion people. *Science* **2010**, *327*, 812–818. [CrossRef]
2. Foley, J.A.; Ramankutty, N.; Brauman, K.A.; Cassidy, E.S.; Gerber, J.S.; Johnston, M.; Mueller, N.D.; O'Connell, C.; Ray, D.K.; West, P.C.; et al. Solutions for a cultivated planet. *Nature* **2011**, *478*, 337–342. [CrossRef]
3. Ray, D.K.; Mueller, N.D.; West, P.C.; Foley, J.A. Yield trends are insufficient to double global crop production by 2050. *PLoS ONE* **2013**, *8*, e66428. [CrossRef]
4. Fischer, E.M.; Sedláček, J.; Hawkins, E.; Knutti, R. Models agree on forced response pattern of precipitation and temperature extremes. *Geophys. Res. Lett.* **2014**, *41*, 8554–8562. [CrossRef]
5. Islam, S.M.F.; Karim, Z. World's demand for food and water: The consequences of climate change. In *Desalination-Challenges and Opportunities*; Farahani, M.H.D.A., Vatanpour, V., Taheri, A., Eds.; IntechOPen: London, UK, 2019; pp. 225–240.
6. Shiferaw, B.; Smale, M.; Braun, H.J.; Duveiller, E.; Reynolds, M.; Muricho, G. Crops that feed the world 10. Past successes and future challenges to the role played by wheat in global food security. *Food Secur.* **2013**, *5*, 291–317. [CrossRef]
7. Bernas, J.; Koppensteiner, L.J.; Tichá, M.; Kaul, H.P.; Klimek-Kopyra, A.; Euteneuer, P.; Moitzi, G.; Neugschwandtner, R.W. Optimal environmental design of nitrogen application rate for facultative wheat using life cycle assessment. *Eur. J. Agron.* **2023**, *146*, 126813. [CrossRef]
8. Jaskulska, I.; Jaskulski, D.; Gałęzewski, L.; Knapowski, T.; Kozera, W.; Wacławowicz, R. Mineral composition and baking value of winter wheat grain under varied environmental and agronomic conditions. *J. Chem.* **2018**, *2018*, 5013825. [CrossRef]
9. Jarecki, W.; Czernicka, M. Yield and quality of winter wheat (*Triticum aestivum* L.) depending on multi-component foliar fertilization. *J. Elem.* **2022**, *27*, 559–567. [CrossRef]
10. Jarecki, W. Effects of sowing date variation on winter wheat (*Tritium aestivum* L.) quality and grain yield. *J. Elem.* **2023**, *28*, 1089–1100. [CrossRef]
11. Šramková, Z.; Gregova, E.; Šturdik, E. Chemical composition and nutritional quality of wheat grain. *Acta Chim. Slovaca* **2009**, *2*, 115–138.
12. Al-Saleh, A.; Brennan, C.S. Bread wheat quality: Some physical, chemical and rheological characteristics of Syrian and English bread wheat samples. *Foods* **2012**, *1*, 3–17. [CrossRef] [PubMed]
13. Budzyński, W. Common wheat. In *Wheats—Common, Spelt, Durum. Cultivation and Uses*; Budzyński, W., Ed.; PWRiL: Poznań, Poland, 2012; 328p. (In Polish)
14. Barak, S.; Mudgil, D.; Khatkar, B.S. Biochemical and functional properties of wheat gliadins: A review. *Crit. Rev. Food Sci. Nutr.* **2015**, *55*, 357–368. [CrossRef] [PubMed]
15. Podolska, G.; Aleksandrowicz, E.; Szafrańska, A. Bread making potential of *Triticum aestivum* and *Triticum spelta* species. *Open Life Sci.* **2020**, *15*, 30–40. [CrossRef]
16. Rothkaehl, J. Wheat grain for human consumption—Marketing and processing. In *Wheats—Common, Spelt, Durum. Cultivation and Uses*; Budzyński, W., Ed.; PWRiL: Poznań, Poland, 2012; pp. 235–264. (In Polish)
17. Zhai, S.; Liu, J.; Xu, D.; Wen, W.; Yan, J.; Zhang, P.; Wan, Y.; Cao, S.; Hao, Y.; Xia, X. A Genome-wide association study reveals a rich genetic architecture of flour color-related traits in bread wheat. *Front. Plant Sci.* **2018**, *9*, 1136. [CrossRef] [PubMed]
18. Rodrighero, M.B.; Caires, E.F.; Lopes, R.B.; Zielinski, A.A.; Granato, D.; Demiate, I.M. Wheat technological quality as affected by nitrogen fertilization under a no-till system. *Acta Sci. Technol.* **2015**, *37*, 175–181. [CrossRef]
19. Yildirim, A.; Atasoy, A. Quality characteristics of some durum wheat varieties grown in the Southeastern Anatolia Region of Turkey (GAP). *Harran J. Agric. Food Sci.* **2020**, *24*, 420–431. [CrossRef]
20. Wei, Y. *Cereal and Food Quality*; Shanxi People Press: Taiyuan, China, 2002.
21. Kulkarni, S.D.; Acharya, R.; Nair, A.G.C.; Rajurkar, N.S.; Reddy, A.V.R. Determination of elemental concentration profiles in tender wheatgrass (*Triticum aestivum* L.) using instrumental neutron activation analysis. *Food Chem.* **2006**, *95*, 699–707. [CrossRef]
22. Piironen, V.; Salmenkallio-Marttila, M. Micronutrients and phytochemicals in wheat grain. In *Wheat: Chemistry and Technology*; American Association of Cereal Chemists: St Paul, MN, USA, 2009; pp. 179–222.
23. Czaja, T.; Sobota, A.; Szostak, R. Quantification of ash and moisture in wheat flour by Raman spectroscopy. *Foods* **2020**, *9*, 280. [CrossRef]
24. Zhygunov, D.; Barkovska, Y.; Yehorshyn, Y.; Zhyhunova, H.; Barikian, K. Type 600 wheat-spelt, flour with improved bakery properties. *Food Sci. Technol.* **2020**, *14*, 53–62. [CrossRef]
25. Li, Y.Q.; Zhu, R.J.; Tian, J.C. Influence of wheat protein contents and fractions on dough rheological properties as determined by using a reconstitution method. *Agric. Sci. China* **2008**, *7*, 395–404. [CrossRef]

26. Trevisan, S.; Khorshidi, A.S.; Scanlon, M.G. Relationship between nitrogen functionality and wheat flour dough rheology: Extensional and shear approaches. *Food Res. Int.* **2022**, *162*, 112049. [CrossRef] [PubMed]
27. Cichoń, Z.; Ptak, M. Analysis of the quality of selected wheat flour types. *Zesz. Nauk. Akad. Ekon. Krakowie* **2005**, *678*, 89–102. (In Polish)
28. Simón, M.R.; Fleitas, M.C.; Castro, A.C.; Schierenbeck, M. How foliar fungal diseases affect nitrogen dynamics, milling, and end-use quality of wheat. *Front. Plant Sci.* **2020**, *11*, 569401. [CrossRef]
29. Ellmann, T. Effect of plant protection, nitrogen fertilization and date of harvest on yield of winter wheat. *Fragm. Agron.* **2011**, *28*, 15–25.
30. Fu, B.X.; Wang, K.; Dupuis, B. Predicting water absorption of wheat flour using high shear-based GlutoPeak test. *J. Cereal Sci.* **2017**, *76*, 116–121. [CrossRef]
31. Puhr, D.P.; D'appolonia, B.L. Effect of baking absorption on bread yield, crumb moisture, and crumb water activity. *Cereal Chem.* **1992**, *69*, 582.
32. Guttieri, M.J.; Bowen, D.; Gannon, D.; O'Brien, K.; Souza, E. Solvent retention capacities of irrigated soft white spring wheat flours. *Crop Sci.* **2001**, *41*, 1054–1061. [CrossRef]
33. Li, J.; Zhu, Y.; Yadav, M.P.; Li, J. Effect of various hydrocolloids on the physical and fermentation properties of dough. *Food Chem.* **2019**, *271*, 165–173. [CrossRef]
34. Marchetti, L.; Cardós, M.; Campaña, L.; Ferrero, C. Effect of glutens of different quality on dough characteristics and breadmaking performance. *LWT-Food Sci. Technol.* **2012**, *46*, 224–231. [CrossRef]
35. Amjid, M.R.; Shehzad, A.; Hussain, S.; Shabbir, M.A.; Khan, M.R.; Shoaib, M. A comprehensive review on wheat flour dough rheology. *Pak. J. Food Sci.* **2013**, *23*, 105–123.
36. Aydoğan, S.; Şahin, M.; Akçacık, A. Relationships between farinograph parameters and bread volume, physicochemical traits in bread wheat flours. *Crop Sci.* **2015**, *3*, 14–18.
37. Menkinoska, M.; Blazhevska, T.; Stamatovsk, V. Determination of rheological properties with farinograf and extensigraf of bio-fortified flour. *Proc. Univ. Ruse* **2018**, *57*, 22–25.
38. Różyło, R.; Laskowski, J.; Dziki, D. Physical properties of wheat bread baked from dough with different parameters. *Acta Agrophys.* **2011**, *18*, 421–430. (In Polish)
39. Dziki, D.; Siastała, M.; Laskowski, J. Changes in physical properties of wheat bread as a result of soy flour addition. *Acta Agrophys.* **2010**, *15*, 91–100. (In Polish)
40. Sahi, S.S.; Little, K.; Ananingsih, V.K. Quality control. In *Bakery Products Science and Technology*; John Wiley & Sons, Ltd.: Chichester, UK, 2014; pp. 489–509.
41. Švec, I.; Hrušková, M. Modelling of wheat, flour, and bread quality parameters. *Sci. Agric. Bohem.* **2009**, *40*, 58–66.
42. Muste, S.; Modoran, C.; Man, S.; Mureşan, V.; Birou, A. The influence of wheat genotype on its quality. *J. Agroaliment. Process. Technol.* **2010**, *16*, 99–103.
43. Jankowski, K.J.; Budzyński, W.S.; Kijewski, Ł.; Dubis, D.; Lemański, M. Flour quality, the rheological properties of dough and the quality of bread made from the grain of winter wheat grown in a continuous cropping system. *Acta Sci. Pol. Agric.* **2014**, *13*, 3–18.
44. Jankowski, K.J.; Kijewski, L.; Dubis, B. Milling quality and flour strength of the grain of winter wheat grown in monoculture. *Rom. Agric. Res.* **2015**, *32*, 191–200.
45. Szafrańska, A.; Cacak-Pietrzak, G.; Sułek, A. Influence of nitrogen fertilization and retardants on the baking value of winter wheat. *Electron. J. Pol. Agric. Univ. EJPAU Agron.* **2008**, *11*, 28.
46. Sułek, A.; Cacak-Pietrzak, G.; Wyzinska, M.; Nieróbca, A. influence of nitrogen fertilization on the yields and grain quality of winter wheat under different environmental conditions. *Int. J. Agric. Biol. Eng.* **2019**, *13*, 127–133.
47. Geisslitz, S.; Longin, C.F.H.; Scherf, K.A.; Koehler, P. Comparative study on gluten protein composition of ancient (einkorn, emmer and spelt) and modern wheat species (durum and common wheat). *Foods* **2019**, *8*, 409. [CrossRef] [PubMed]
48. Andersson, A.; Johansson, E.; Oscarson, P. Nitrogen redistribution from the roots in post-anthesis plants of spring wheat. *Plant Soil* **2005**, *269*, 321–332. [CrossRef]
49. Andersson, A.; Johansson, E. Nitrogen partitioning in entire plants of different spring wheat cultivars. *J. Agron. Crop Sci.* **2006**, *192*, 121–131. [CrossRef]
50. Stankowski, S.; Rutkowska, A. Winter wheat grain and flour quality traits as affected by the nitrogen fertilization dose and date. *Acta Sci. Pol. Agric.* **2006**, *5*, 53–61. (In Polish)
51. Rozbicki, J.; Ceglińska, A.; Gozdowski, D.; Jakubczak, M.; Cacak-Pietrzak, G.; Mądry, W.; Golba, J.; Piechociński, M.; Sobczyński, M.; Studnicki, M.; et al. Influence of the cultivar, environment and management on the grain yield and bread-making quality in winter wheat. *J. Cereal Sci.* **2015**, *61*, 126–132. [CrossRef]
52. Warechowska, M.; Stępień, A.; Wojtkowiak, K.; Nawrocka, A. The impact of nitrogen fertilization strategies on selected qualitative parameters of spring wheat grain and flour. *Pol. J. Nat. Sci.* **2019**, *34*, 199–212.
53. Souza, E.J.; Martin, J.M.; Guttieri, M.J.; O'brien, K.M.; Habernicht, D.K.; Lanning, S.P.; McLean, R.; Carlson, G.R.; Talbert, L.E. Influence of genotype, environment, and nitrogen management on spring wheat quality. *Crop Sci.* **2004**, *44*, 425–432. [CrossRef]
54. Dennett, A.L.; Trethowan, R.M. Milling efficiency of triticale grain for commercial flour production. *J. Cereal Sci.* **2013**, *57*, 527–530. [CrossRef]

55. Joubert, M.; Lullien-Pellerin, V.; Morel, M.H. Impact of durum wheat grain composition on semolina yield and pasta quality. In Proceedings of the 15th European Young Cereal Scientists and Technologists Workshop (EYCSTW), Milan/Bergamo, Italy, 26 April 2016.
56. Souza, T.M.; Prando, A.M.; de Miranda, M.Z.; Hirooka, E.Y.; Zucareli, C. Kernel chemical composition and flour quality of wheat in response to nitrogen sources and doses. *Rev. Agrar.* **2019**, *12*, 528–541. [CrossRef]
57. Siddiqi, R.A.; Singh, T.P.; Rani, M.; Sogi, D.S.; Bhat, M.A. Diversity in grain, flour, amino acid composition, protein profiling, and proportion of total flour proteins of different wheat cultivars of North India. *Front. Nutr.* **2020**, *7*, 141. [CrossRef]
58. Banach, J.K.; Majewska, K.; Żuk-Gołaszewska, K. Effect of cultivation system on quality changes in durum wheat grain and flour produced in North-Eastern Europe. *PLoS ONE* **2021**, *16*, e0236617. [CrossRef] [PubMed]
59. Hong, S.; Park, J.; Lee, G.E.; Yoon, Y.M.; Kang, C.S.; Park, C.S. Effects of nitrogen fertilization on o-free, Korean wheat cultivar reduced ω-5 gliadin, on agronomic traits and noodle properties. *Korean Soc. Breed. Sci.* **2023**, *55*, 126–136. [CrossRef]
60. Saint Pierre, C.; Peterson, C.J.; Ross, A.S.; Ohm, J.B.; Verhoeven, M.C.; Larson, M.; Hoefer, B. White wheat grain quality changes with genotype, nitrogen fertilization, and water stress. *Agron. J.* **2008**, *100*, 414–420. [CrossRef]
61. Valdés Valdés, C.; Estrada-Campuzano, G.; Martinez Rueda, C.G.; Domínguez López, A.; Solis-Moya, E.; Villanueva Carvajal, A. Grain and flour wheat quality modified by genotype, availability of nitrogen, and growing season. *Int. J. Agron.* **2020**, *2020*, 1974083. [CrossRef]
62. Keres, I.; Alaru, M.; Koppel, R.; Altosaar, I.; Tosens, T.; Loit, E. The combined effect of nitrogen treatment and weather conditions on wheat protein-starch interaction and dough quality. *Agriculture* **2021**, *11*, 1232. [CrossRef]
63. Majewska, K. Classification and synthesis of wheat grain gluten proteins. *Żyw. Nauk. Technol. Jakość* **1999**, *6*, 15–25. (In Polish)
64. Blandino, M.; Marinaccio, F.; Reyneri, A. Effect of late-season nitrogen fertilization on grain yield and on flour rheological quality and stability in common wheat, under different production situations. *Ital. J. Agron.* **2016**, *11*, 107–113. [CrossRef]
65. Xue, C.; Matros, A.; Mock, H.P.; Mühling, K.H. Protein composition and baking quality of wheat flour as affected by split nitrogen application. *Front. Plant Sci.* **2019**, *10*, 642. [CrossRef]
66. Guerrini, L.; Napoli, M.; Mancini, M.; Masella, P.; Cappelli, A.; Parenti, A.; Orlandini, S. Wheat grain composition, dough rheology, and bread quality as affected by nitrogen and sulfur fertilization and seeding density. *Agronomy* **2020**, *10*, 233. [CrossRef]
67. Jańczak-Pieniążek, M.; Buczek, J.; Jarecki, W.; Bobrecka-Jamro, D. Effect of high nitrogen doses on yield, quality, and chemical composition of grain of winter wheat cultivars. *J. Elem.* **2020**, *25*, 1005–1017.
68. Cesevičienė, J.; Gorash, A.; Liatukas, Ž.; Armonienė, R.; Ruzgas, V.; Statkevičiūtė, G.; Jaškūnė, K.; Brazauskas, G. Grain yield performance and quality characteristics of waxy and non-waxy winter wheat cultivars under high and low-input farming systems. *Plants* **2022**, *11*, 882. [CrossRef]
69. Zende, N.B.; Sethi, H.N.; Karunakar, A.P.; Jiotode, D.J. Effect of sowing time and fertility levels on yield and quality of durum wheat genotypes. *Res. Crops* **2005**, *6*, 194–196.
70. Otteson, B.N.; Mergoum, M.; Ransom, J.K. Seeding rate and nitrogen management on milling and baking quality of hard red spring wheat genotypes. *Crop Sci.* **2008**, *48*, 749–755. [CrossRef]
71. Caglar, O.; Bulut, S.; Karaoglu, M.M.; Kotancilar, H.G.; Ozturk, A. Quality response of facultative wheat to winter sowing, freezing sowing and spring sowing at different seeding rates. *J. Anim. Vet. Adv.* **2011**, *10*, 3368–3374. [CrossRef]
72. Zecevic, V.; Boskovic, J.; Knezevic, D.; Micanovic, D. Effect of seeding rate on grain quality of winter wheat. *Chil. J. Agric. Res.* **2014**, *74*, 23–28. [CrossRef]
73. Soofizada, Q.; Pescatore, A.; Guerrini, L.; Fabbri, C.; Mancini, M.; Orlandini, S.; Napoli, M. Effects of nitrogen plus sulfur fertilization and seeding density on yield, rheological parameters, and asparagine content in old varieties of common wheat (*Triticum aestivum* L.). *Agronomy* **2022**, *12*, 351. [CrossRef]
74. Wilcox, J.; Makowski, D. A meta-analysis of the predicted effects of climate change on wheat yields using simulation studies. *Field Crops Res.* **2014**, *156*, 180–190. [CrossRef]
75. Pakrooh, P.; Kamal, M.A. Modeling the potential impacts of climate change on wheat yield in Iran: Evidence from national and provincial data analysis. *Ecol. Model.* **2023**, *486*, 110513. [CrossRef]
76. Farooq, A.; Farooq, N.; Akbar, H.; Hassan, Z.U.; Gheewala, S.H. A critical review of climate change impact at a global scale on cereal crop production. *Agronomy* **2023**, *13*, 162. [CrossRef]
77. Zahra, N.; Hafeez, M.B.; Wahid, A.; Al Masruri, M.H.; Ullah, A.; Siddique, K.H.; Farooq, M. Impact of climate change on wheat grain composition and quality. *J. Sci. Food Agric.* **2023**, *103*, 2745–2751. [CrossRef]
78. Nóia Júnior, R.d.S.; Deswarte, J.-C.; Cohan, J.-P.; Martre, P.; van Der Velde, M.; Lecerf, R.; Webber, H.; Ewert, F.; Ruane, A.C.; Slafer, G.A.; et al. The extreme 2016 wheat yield failure in France. *Glob. Chang. Biol.* **2023**, *29*, 3130–3146. [CrossRef] [PubMed]
79. Ainsworth, E.A.; Ort, D.R. How do we improve crop production in a warming world? *Plant Physiol.* **2010**, *154*, 526–530. [CrossRef] [PubMed]
80. Xiao, D.; Tao, F.; Liu, Y.; Shi, W.; Wang, M.; Liu, F.; Shuai, Z.; Zhu, Z. Observed changes in winter wheat phenology in the North China Plain for 1981–2009. *Int. J. Biometeorol.* **2013**, *57*, 275–285. [CrossRef] [PubMed]
81. Xiao, D.; Moiwo, J.P.; Tao, F.; Yang, Y.; Shen, Y.; Xu, Q.; Liu, J.; Liu, F. Spatiotemporal variability of winter wheat phenology in response to weather and climate variability in China. *Mitig. Adapt. Strateg. Glob. Chang.* **2015**, *20*, 1191–1202. [CrossRef]

82. Kusunose, Y.; Rossi, J.J.; van Sanford, D.A.; Alderman, F.D.; Anderson, J.A.; Chai, Y.; Gerullis, M.K.; Jagadish, S.V.K.; Paul, P.A.; Tack, S.B.; et al. Sustaining productivity gains in the face of climate change: A research agenda for US wheat. *Glob. Chang. Biol.* **2023**, *29*, 926–934. [CrossRef]
83. Qiao, S.; Harrison, S.P.; Prentice, I.C.; Wang, H. Optimality-based modelling of wheat sowing dates globally. *Agric. Syst.* **2023**, *206*, 103608. [CrossRef]
84. Alignan, M.; Roche, J.; Bouniols, A.; Cerny, M.; Mouloungui, Z.; Merah, O. Effects of genotype and sowing date on phytostanol–phytosterol content and agronomic traits in wheat under organic agriculture. *Food Chem.* **2009**, *117*, 219–225. [CrossRef]
85. Ahmed, M.; Hassan, F. Response of spring wheat (*Triticum aestivum* L.) quality traits and yield to sowing date. *PLoS ONE* **2015**, *10*, e0126097. [PubMed]
86. Guzmán, C.; Autrique, J.E.; Mondal, S.; Singh, R.P.; Govindan, V.; Morales-Dorantes, A.; Posadas-Romano, G.; Crossa, J.; Ammar, K.; Peña, R.J. Response to drought and heat stress on wheat quality, with special emphasis on bread-making quality, in durum wheat. *Field Crops Res.* **2016**, *186*, 157–165. [CrossRef]
87. Tahir, I.S.; Nakata, N.; Ali, A.M.; Mustafa, H.M.; Saad, A.S.I.; Takata, K.; Ishikawa, N.; Abdalla, O.S. Genotypic and temperature effects on wheat grain yield and quality in a hot irrigated environment. *Plant Breed.* **2006**, *125*, 323–330. [CrossRef]
88. Li, Y.F.; Wu, Y.; Hernandez-Espinosa, N.; Peña, R.J. Heat and drought stress on durum wheat: Responses of genotypes, yield, and quality parameters. *J. Cereal Sci.* **2013**, *57*, 398–404. [CrossRef]
89. Johansson, E.; Kuktaite, R.; Andersson, A.; Prieto-Linde, M.L. Protein polymer build-up during wheat grain development: Influences of temperature and nitrogen timing. *J. Sci. Food Agric.* **2005**, *85*, 473–479. [CrossRef]
90. Labuschagne, M.T.; Elago, O.; Koen, E. The influence of temperature extremes on some quality and starch characteristics in bread, biscuit and durum wheat. *J. Cereal Sci.* **2009**, *9*, 84–189. [CrossRef]
91. Hrušková, M.; Švec, I. Wheat hardness in relation to other quality factors. *Czech J. Food Sci.* **2009**, *27*, 240–248. [CrossRef]
92. Knapowski, T.; Ropińska, P.; Kazek, M.; Wenda-Piesik, A. Flour and bread quality of spring wheat cultivars (*Triticum aestivum* L.) sown at facultative and spring sowing dates. *Acta Sci. Pol. Agric.* **2018**, *17*, 133–142.
93. Dong, S.; Zhang, X.; Chu, J.; Zheng, F.; Fei, L.; Dai, X.; He, M. Optimized seeding rate and nitrogen topdressing ratio for simultaneous improvement of grain yield and bread-making quality in bread wheat sown on different dates. *J. Sci. Food Agric.* **2021**, *102*, 360–369. [CrossRef]
94. Lachutta, K.; Jankowski, K.J. An agronomic efficiency analysis of winter wheat at different sowing strategies and nitrogen fertilizer rates: A case study in northeastern Poland. *Agriculture* **2024**, *14*, 442. [CrossRef]
95. Lachutta, K.; Jankowski, K.J. The quality of winter wheat grain at different sowing strategies and nitrogen fertilizer rates: A case study in northeastern Poland. *Agriculture* **2024**, *14*, 552. [CrossRef]
96. *PN-EN ISO 5530-1:2015-01*; Wheat Flour—Physical Characteristics of Doughs. Part 1. Determination of Water Absorption and Rheological Properties Using a Farinograph. Polish Committee for Standardization: Warszawa, Poland, 2015. (In Polish)
97. Klockiewicz-Kamińska, E.; Brzeziński, W.J. Method of evaluation and quality-based classification of cultivars. *Wiadomości Odmianozn.* **1997**, *67*, 3–18. (In Polish)
98. TIBCO Software Inc. *Statistica (Data Analysis Software System), Version 13*; TIBCO Software Inc.: Palo Alto, CA, USA, 2017.
99. Ozturk, A.; Aydin, F. Effect of water stress at various growth stages on some quality characteristics of winter wheat. *J. Agron. Crop Sci.* **2004**, *190*, 93–99. [CrossRef]
100. Egesel, C.O.; Kahriman, F.S.T.L.; Baytekin, H. Interrelationships of flour quality traits with grain yield in bread wheat and choosing suitable cultivars. *Anadolu J. Agric. Sci.* **2009**, *24*, 76–83.
101. Adeel, M.; Nazir, A.; Aziz, H. Impact of sowing dates and terminal heat stress on wheat grain flour physical and chemical properties. *Jammu Kashmir J. Agric.* **2022**, *2*, 25–34. [CrossRef]
102. Garrido-Lestache, E.; López-Bellido, R.J.; López-Bellido, L. Effect of N rate, timing, and splitting and N type on bread-making quality in hard red spring wheat under rainfed Mediterranean conditions. *Field Crops Res.* **2004**, *85*, 213–236. [CrossRef]
103. Munsif, F.; Arif, M.; Ali, K.; Jan, M.T.; Khan, M.J. Influence of planting dates on grain quality of different wheat cultivars in dual-purpose system. *Int. J. Agric. Biol.* **2015**, *17*, 945–952. [CrossRef]
104. Biel, W.; Stankowski, S.; Sobolewska, M.; Sadkiewicz, J.; Jaroszewska, A.; Pużyński, S. Effect of selected agronomic factors on the baking quality of winter spelt strains and cultivars (*Triticum aestivum* ssp. *spelta* L.) in comparison with common wheat (*Triticum aestivum* ssp. *vulgare*). *Rom. Agric. Res.* **2016**, *33*, 251–258.
105. Caglar, O.; Yildiz, G.; Karaoglu, M.M.; Ozturk, A.; Bulut, S. The effects of sowing times and seeding rates on the farinograph parameters and color of facultative kirik wheat. *Fresenius Environ. Bull.* **2023**, *32*, 2918–2924.
106. Wieser, H. Chemistry of gluten proteins. *Food Microbiol.* **2007**, *24*, 15–119. [CrossRef]
107. Bagulho, A.S.; Costa, R.; Almeida, A.S.; Pinheiro, N.; Moreira, J.; Gomes, C.; Coco, J.; Costa, A.; Coutinho, J.; Maçãs, B. Influence of year and sowing date on bread wheat quality under Mediterranean conditions. *Emir. J. Food Agric.* **2015**, *27*, 186–199. [CrossRef]
108. Balla, K.; Veisz, O. Changes in the quality of cereals in response to heat and drought stress. *Acta Agron. Óvar.* **2007**, *49*, 451–455.
109. Geleta, B.; Atak, M.; Baenziger, P.S.; Nelson, L.A.; Baltenesperger, D.D.; Eskridge, K.M.; Shipman, M.J.; Shelton, D.R. Seeding rate and genotype effect on agronomic performance and end-use quality of winter wheat. *Crop Sci.* **2002**, *42*, 827–832.

110. Mikos-Szymańska, M.; Podolska, G. The effects of sowing date and seeding rate on spelt and common wheat protein composition and characteristics. *Qual. Assur. Saf. Crops Foods* **2016**, *8*, 289–300. [CrossRef]
111. Hao, R.; Noor, H.; Wang, P.; Sun, M.; Noor, F.; Ullah, S.; Gao, Z. Combined effects of starch sucrose content and planting density on grain protein content of winter wheat (*Triticum aestivum* L.). *J. Food Nutr. Res.* **2022**, *10*, 321–331.
112. Madan, H.S.; Munjal, R. Effect of split doses of nitrogen and seed rate on protein content, protein fractions and yield of wheat. *J. Agric. Biol. Sci.* **2009**, *4*, 26–31.
113. Johansson, E.; Prieto-Linde, M.L.; Svensson, G. Influence of nitrogen application rate and timing on grain protein composition and gluten strength in Swedish wheat cultivars. *J. Plant Nutr. Soil Sci.* **2004**, *167*, 345–350. [CrossRef]
114. Rossmann, A.; Pitann, B.; Mühling, K.H. Splitting nitrogen applications improves wheat storage protein composition under low N supply. *J. Plant Nutr. Soil Sci.* **2019**, *182*, 347–355. [CrossRef]
115. Luo, C.; Branlard, G.; Griffin, W.B.; McNeil, D.L. The effect of nitrogen and sulphur fertilisation and their interaction with genotype on wheat glutenins and quality parameters. *J. Cereal Sci.* **2000**, *31*, 185–194. [CrossRef]
116. Haile, D.; Nigussie, D.; Ayana, A. Nitrogen use efficiency of bread wheat: Effects of nitrogen rate and time of application. *J. Soil Sci. Plant Nutr.* **2012**, *12*, 389–410.
117. Linlaud, N.E.; Puppo, M.C.; Ferrero, C. Effect of hydrocolloids on water absorption of wheat flour and farinograph and textural characteristics of dough. *Cereal Chem.* **2009**, *86*, 376–382. [CrossRef]
118. Sapirstein, H.; Wu, Y.; Koksel, F.; Graf, R. A study of factors influencing the water absorption capacity of Canadian hard red winter wheat flour. *J. Cereal Sci.* **2018**, *81*, 52–59. [CrossRef]
119. Zhang, Z.; Jia, D.; Wang, D.; Zhou, N.; Xing, Z.; Xu, K.; Wei, H.; Guo, B.; Zhang, H. Starch and dough-related properties of wheat (*Triticum aestivum* L.) exposed to varying temperatures and radiances after anthesis. *Agronomy* **2023**, *13*, 1069. [CrossRef]
120. Maningat, C.C.; Seib, P.A.; Bassi, S.D.; Woo, K.S.; Lasater, G.D. Wheat starch: Production, properties, modification and uses. In *Starch*; Academic Press: Cambridge, MA, USA, 2009; pp. 441–510.
121. Huang, X.; Zhou, X.; Liu, X.; Zhong, W.; Wang, X.; Ju, Z.; Yin, Y.; Xin, Q.; Liu, N.; Liu, X.; et al. Structural and physicochemical effects on the starch quality of the high-quality wheat genotype caused by delayed sowing. *Front. Nutr.* **2024**, *11*, 1389745. [CrossRef] [PubMed]
122. Zang, Y.; Yao, H.; Ran, L.; Zhang, R.; Duan, Y.; Yu, X.; Xiong, F. Physicochemical properties of wheat starch under different sowing dates. *Starch-Stärke* **2022**, *74*, 2100290. [CrossRef]
123. Chen, Y.; Yang, P.; Li, H. Simulation study on effects of planting density and sowing date on grain quality of winter wheat. *J. Agric. Sci. Technol.* **2023**, *24*, 143–153.
124. Han, H.; Yang, W. Influence of uniconazole and plant density on nitrogen content and grain quality in winter wheat in South China. *Plant Soil Environ.* **2009**, *55*, 159–166. [CrossRef]
125. de Pelegrin, A.J.; Carvalho, I.R.; Ferrari, M.; Nardino, M.; Szareski, V.J.; Meira, D.; Wartha, C.A.; Follman, D.N.; de Pelegrin, C.M.G.; Gutkoski, L.C.; et al. Evaluation of solvent retention capacity of wheat (*Triticum aestivum* L.) flour depending on genotype and different timing of nitrogenous fertilizer application. *Afr. J. Agric. Res.* **2016**, *11*, 4389–4394.
126. Gawęda, D.; Gawęda, M.; Chojnacka, S.; Sobolewska, M.; Łukasz, J.; Hury, G.; Wesołowska-Trojanowska, M. Evaluation of the technological quality of grain and flour of two spelt wheat (*Triticum aestivum* ssp. *spelta* L.) cultivars grown under different conditions of crop protection and seeding rate. *Appl. Ecol. Environ. Res.* **2019**, *17*, 4377–4395. [CrossRef]
127. Nakano, H.; Morita, S. Effects of seeding rate and nitrogen application rate on grain yield and protein content of the bread wheat cultivar 'Minaminokaori' in southwestern Japan. *Plant Prod. Sci.* **2009**, *12*, 109–115. [CrossRef]
128. Gaile, Z.; Ruza, A.; Kreita, D.; Paura, L. Yield components and quality parameters of winter wheat depending on tillering coefficient. *Agron. Res.* **2017**, *15*, 79–93.
129. Mosanaei, H.; Ajamnorozi, H.; Dadashi, M.R.; Faraji, A.; Pessarakli, M. Improvement effect of nitrogen fertilizer and plant density on wheat (*Triticum aestivum* L.) seed deterioration and yield. *Emir. J. Food Agric.* **2017**, *29*, 899–910. [CrossRef]
130. Shevkani, K.; Singh, N.; Bajaj, R.; Kaur, A. Wheat starch production, structure, functionality and applications—A review. *Int. J. Food Sci. Technol.* **2017**, *52*, 38–58. [CrossRef]
131. Zhang, Y.; Zhang, G. Starch content and physicochemical properties of green wheat starch. *Int. J. Food Prop.* **2019**, *22*, 1463–1474. [CrossRef]
132. Zhang, Y.; Dai, X.; Jia, D.; Li, H.; Wang, Y.; Li, C.; Xu, H.; He, M. Effects of plant density on grain yield, protein size distribution, and breadmaking quality of winter wheat grown under two nitrogen fertilisation rates. *Eur. J. Agron.* **2016**, *73*, 1–10. [CrossRef]
133. Fleitas, M.C.; Schierenbeck, M.; Gerard, G.S.; Dietz, J.I.; Golik, S.I.; Campos, P.E.; Simón, M.R. How leaf rust disease and its control with fungicides affect dough properties, gluten quality and loaf volume under different N rates in wheat. *J. Cereal Sci.* **2018**, *80*, 119–127. [CrossRef]
134. Kunkulberga, D.; Linina, A.; Ruza, A. Effect of nitrogen fertilization on protein content and rheological properties of winter wheat wholemeal. In Proceedings of the 3th Baltic Conference on Food Science and Technology "FOOD, NUTRITION, WELL-BEING", Jelgava, Latvia, 2–3 May 2019; pp. 88–92.
135. Lagrain, B.; Wilderjans, E.; Glorieux, C.; Delcour, J.A. Importance of gluten and starch for structural and textural properties of crumb from fresh and stored bread. *Food Biophys.* **2012**, *7*, 173–181. [CrossRef]

136. Knapowski, T.; Ralcewicz, M. Evaluation of qualitative features of Mikon cultivar winter wheat grain and flour depending on selected agronomic factors. *Electron. J. Pol. Agric. Univ. EJPAU Agron.* **2004**, *7*, 1–12.
137. Knapowski, T.; Ralcewicz, M. Estimation of the quality parameters of winter wheat grain and flour in response to different nitrogen rates. *Ann. UMCS Sec. E* **2004**, *59*, 959–968. (In Polish)

applied sciences

MDPI

Article

Optimum Transplanting Date for Rape Forage and Grain Yields in the Ridge Culture Place Planting System on the Yangtze River Delta

Yueyue Tao [1,2,*], Dongmei Li [1,2], Yiwen Yu [1,3], Changying Lu [1,2], Meng Huang [1,3], Haihou Wang [1,2] and Hua Sun [1,3,*]

1 Institute of Agricultural Sciences in Taihu Lake District, Suzhou Academy of Agricultural Sciences, Suzhou 215155, China; sznkyldm@163.com (D.L.); yywsznky2024@163.com (Y.Y.); luchangying@163.com (C.L.); sznkyhm@163.com (M.H.); wanghaihou@126.com (H.W.)

2 National Soil Quality Observation and Experimental Station in Xiangcheng, Suzhou 215155, China

3 Suzhou Comprehensive Experimental Station of China Rape Research System, Suzhou 215155, China

* Correspondence: twhhltyy@163.com (Y.T.); sunhqzy@163.com (H.S.)

Abstract: The ridge culture place planting system (RCPPS) is a promising technique for planting rapeseed that can promote the growth of rapeseed by late rice stubble, which has been widely adopted in the Yangtze River delta. To determine the optimum planting date for rape (*Brassica napus* L.) forage and grain yield in an intensive rice–rape rotation system, a field experiment was conducted with five transplantation dates (from 20 October to 30 November at 10 day intervals) in RCPPS. The forage/grain yield, nutrition, and growth parameters were analyzed. At podding, rape biomass yield was highest, and no significant differences were found among treatments. It was around 12.0% crude protein, 11.4% ether extract, 38.8% neutral detergent fiber, and 34.9% acid detergent fiber. In the treatments of 20 and 30 November, crude protein content increased and acid detergent fiber content decreased significantly. Compared with 20 October, the grain yield of rape transplanted in November decreased significantly by 17.2% to 22.5%. The grain yield was significantly correlated with the number of secondary branches, pods, and seeds. At the final flowering stage, rape transplanted in November had noticeably reduced leaf growth, rhizome width, and yield than 20 and 30 October. Overall, for multiple uses of rapeseed in the Yangtze River delta belt with RCPPS, it is optimal to plant in mid to late November for forage use with higher nutritional value, being coordinated with the previous rice crop, whereas late October is the appropriate planting time to obtain a higher grain yield.

Keywords: ridge culture place planting system; optimum transplant date; rape forage yield; rape grain yield; rape growth characteristics; rape forage nutrition

1. Introduction

The summer rice–winter rapeseed (*Brassica napus* L.) rotation is the traditional cultivation system in the Yangtze River delta, in which the rice crop cycle is from June to October and afterwards, the winter rape typically grows from November to May. It is likely that oilseed rapeseed would increase rice production and improve soil fertility when used as a previous crop in a rotation system [1]. However, considering the late harvest of the preceding crop and low planting profit, traditional rapeseed is often classified as an uncertain crop compared with the local main crop of rice [2]. Therefore, because of the delayed planting date of rapeseed, low air temperature and high precipitation always occur, which inhibit the vegetative growth of rapeseed seedlings and lead to the poor development of pod formation and grain filling in the late period [3,4]. The planting costs of rapeseed have also increased substantially because the price of agricultural production material and labor resources has increased rapidly in the Yangtze River delta. In addition, large quantities of low-cost, high-quality imported oilseeds have negatively affected the development of domestic rapeseed production, in which rapeseed is grown conventionally for grain [5,6].

The hybrid varieties of low erucic acid and glusosiolate ('double-low') rape are high-yielding forage crops characterized by their high regrowth capacity [7] and nutritive value [8]. Thus, forage rape can be a substantial part of the feed rotation in pasture, particularly in autumn–winter when pasture growth is limited, and it therefore increases the economic benefits of rapeseed production. Forage rape has been promoted in a large area in northern China, including in Heilongjiang, Shanxi, and Gansu provinces [9]. In the northern region, forage rape typically grows from August to October, which is beneficial for improving the productivity in the complementary forage rotation of wheat–forage rape or bean–forage rape. In addition to enriching the species of forage in these agricultural areas, forage rape improves resource use efficiency, including that of temperature, water, and light, compared with fallow fields [10,11]. Forage rape is also beneficial because it reduces soil erosion and soil pathogens and improves soil characteristics [7,12].

In 2017, the National Agricultural Technology Extension Service Center issued 'Winter Rapeseed Production Technical Guidance Opinions 2017–2018', which first clarified the multiple uses of rape for oil, forage, flowers, honey, and fertilizer [13]. To improve rapeseed production, innovative uses of rape need to be developed, consistent with local conditions. In the loess hilly region, the combined uses of rapeseed for forage, sightseeing, and grain have promoted the development of multifunctional agriculture, which brings great economic and ecological benefits [9]. In the cold region of Northeast China, the first innovation in growing forage rapeseed was to plant two crops a year [14]. However, few studies have examined the cultivation of forage rape in the Yangtze River delta.

In the rice–rape rotation system in the Yangtze River delta, the planting date of rapes is restricted by various factors such as stubble configuration, weather, and soil conditions. The timing of transplanting inevitably affects the growth characteristics of rapes. Delayed sowing date with insufficient accumulated temperature would directly inhibit seed germination and early seedling growth, thus affecting the accumulation and transport of photosynthetic products, leading to low biomass and grain yield [15,16]. The ridge culture place planting system (RCPPS) is a promising technique for planting rapeseed that has been widely adopted in the Yangtze River delta [17]. With the RCPPS, the stress of waterlogging damage can be avoided, especially in the seedling stages. Furthermore, early seed sowing and seedling raising can facilitate the growth of rapeseed, particularly alleviating the inhibition of rape growth by rice stubble [17,18]. Sun et al. (2015) [17] found the grain yield with the RCPPS is significantly higher than that with traditional transplanting or a direct-seeding system in the Yangtze River delta.

However, information on the effects of transplanting dates on the growths and nutritional value characteristics, as well as the optimum transplanting dates of brassicas for forage and grains with the RCPPS is lacking, which has inhibited adoption by producers. Therefore, in a field experiment with the RCPPS, we compared the effects of different transplanting dates on the characteristics of rape that are important for forage and grain use. The objectives were the following: (1) to quantify and compare the effects of five transplanting dates on yield and nutritional value of forage grass as well as grain yield of rapeseed in the RCPPS, and (2) to identify the appropriate transplanting dates for rape intended for use as forage or grains. The results of this study will provide the basis for the efficient cultivation of rapeseed for multiple uses in the middle and lower reaches of the Yangtze River.

2. Materials and Methods

2.1. Site Description

A field experiment was conducted near Suzhou (31°27′ N, 120°25′ E; 2 m a.s.l.) in Jiangsu Province, China. In September 2016, soil samples were collected at the 0 to 20 cm depth with a soil auger (3.5 cm diameter) from five individual locations in each plot and then mixed together as one replicate. Soil organic carbon (SOC) and total nitrogen (TN) content were determined by using the Walkley–Black and Kjeldahl digestion methods, respectively [19,20]. Soil water-soluble phosphorus (P) was determined by the Olsen method [21]. Soil available potassium (K) was extracted with 1 M NH_4OAC (pH = 7) and analyzed by a flame photometer (M410, Sherwood, UK). Soil pH was measured in a 1:2.5 soil/water solution using a combined electrode pH meter (HI 98121, Hanna Instruments, Kehl am Rhein, Germany). The soil pH was 6.2, and the nutrient concentrations were the following: SOC, 31.6 g·kg^{-1}; TN, 1.80 g·kg^{-1}; available P, 35.0 mg·kg^{-1}; and available K, 137.3 mg·kg^{-1}.

The climate data during the growth period were collected at a meteorological station. Average monthly rainfall and daily temperature are summarized in Table 1. A north subtropical monsoon climate characterizes the Yangtze River delta, with an annual mean air temperature of 16.0 ± 1.9 °C. During the experimental years, the annual average precipitation was approximately 1074 mm, mainly distributed in the beginning of the growing season from October to November. The cumulative annual temperatures reached approximately 3987 °C. The total sunshine hours were 1101.3 h. The relative humidity was 71.9%. The 'double-low' hybrid rapeseed in the experiment was Ningza No. 1818, which was from the Jiangsu Academy of Agricultural Sciences.

Table 1. Meteorological conditions during the rapeseed growing season.

Month	Average Temperature (°C)	Cumulative Temperatures (°C)	Total Rainfall (mm)	Total Sunshine Hours (h)	Relative Air Humidity (%)
October	20.2	615.5	282.9	42.6	82.9
November	13.1	385.2	131.8	87.0	79.6
December	8.9	276.2	55.6	120.5	74.3
January	7.0	199.2	67.3	104.5	74.2
February	7.9	196.7	28.9	130.7	65.4
March	12.8	332.4	74.1	129.2	66.6
April	20.8	538.9	90.4	169.6	63.2
May	26.3	703.0	83.0	186.9	65.0
Average	16.0	3987.0	1074.0	1101.3	71.9

2.2. *Experimental Design and Field Management*

The field experiment was conducted from 2016 to 2017 and was composed of five transplanting dates as experimental treatments that were arranged in a randomized block design with three replications. The five transplanting dates were 20 October, 30 October, 10 November, 20 November, and 30 November, with a 10 day interval, according to the harvest dates of previous rice variety. The size of each plot was 15 m^2 (3 m × 5 m).

The rapeseed was sown in the field in late September and then nursed for approximately 45 days. The seedlings of all transplant date treatments were planted in the RCPPS (Appendix A, Figure A1). The RCPPS in the plots consisted of 80 cm wide ridges of soil with 30 cm wide and 20 cm deep furrows on each side of a ridge. Each plot contained three ridges, and in each ridge, two rows of rapeseed were transplanted. One rape seedling at the age of 4.1 to 4.5 leaves was selected to transplant per hill at constant spacing (15.0 cm × 60.0 cm and 15.0 cm × 40.0 cm) with the transplanting density of 12,000 seedlings·ha^{-1}. Nitrogen fertilizer of chemical urea was applied at 270 kg N ha^{-1} during the entire growth season at the ratio of 4:3:3 at transplanting, seedling, and bolting stages. All treatments were supplied with P (105 kg P_2O_5 ha^{-1}) and K (225 kg K_2O ha^{-1}) as a single application. The other agricultural management practices were consistent with those of local production.

2.3. *Field Sampling and Laboratory Analysis*

Observations of leaf age and growth stages were recorded at 5 day and 7 day intervals, respectively. Biomass yield was measured at the stages of budding, initial flowering, final flowering, and podding. On each sampling date, six hills were harvested. The plants were cut from the bottom, and the fresh weights were recorded. Then, the growth characteristics, namely, the number of effective leaves, rhizome thickness, and plant height, were measured in the lab. Subsamples of approximately 800 g plant were air-dried and then oven-dried at 70 °C until constant weight to determine the dry weight (DM) and water content.

At maturity, 8 m^2 of rapeseed was manually harvested (non-sampling area) from each plot to determine the grain yield. The grains were threshed, cleaned, and weighed after drying in the sun for one week. Subsamples of 500 g of grains were oven-dried at 70 °C until constant weight to determine the DM and water content. To prevent the loss of the pods, before the final harvest, six hills of representative plants were selected randomly to determine the yield components of rapeseed: the number of branches per plant, the number of pods and seeds at each level of branch, and the 1000 seed weight. In addition, subsamples of pods were taken randomly to determine the length and width and the number of seeds.

All plant samples were ground in a micro hammer (FZ102, Taisite, Tianjin, China). To avoid cross-contamination, it was cleaned between millings. After wet digestion with H_2SO_4–H_2O_2 at 220 °C, samples were analyzed for crude protein (CP%) by the micro-Kjeldahl procedure [22]. The contents of neutral detergent fiber (NDF) and acid detergent fiber (ADF) were determined by the Van Soest method [23] The ether extract (EE) was measured by the residual method [24].

2.4. Statistical Analyses

Statistical analyses were conducted using the Statistical Analysis System 9.3 [25] (SAS Institute, Cary, NC, USA, 2009). A one-way analysis of variance (ANOVA) with Tukey's post-hoc test ($p < 0.05$) was used to assess the differences in biomass and grain yield, as well as the parameters of nutritional content, yield components and growth between transplanting date treatments. A Pearson correlation between grain yield and yield components was adapted using SAS. All figures were plotted with Origin 2018 software (OriginLab, Northampton, MA, USA).

3. Results

3.1. Effects of Transplant Date on Forage Yield and Nutrition of Rapeseed

The dry matter production of forage rape increased rapidly from the bud period to the final flowering stage but then increased gradually in the podding stage (Figure 1). At the final flowering stage, the transplantation on 20 October resulted in significantly higher biomass yield than that in the transplantations in November. However, no significant differences were found between transplanting periods in the other stages (Figure 1). Notably, compared with the transplanting dates of 20 and 30 October and 10 November, the 20 and 30 November transplanting treatments had significantly higher CP content and lower ADF content (Figure 2). There were no significant differences in EE and NDF contents among the treatments (Figure 2).

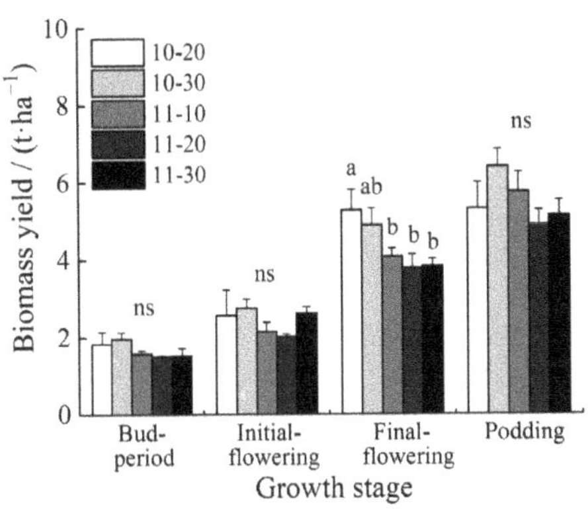

Figure 1. Biomass yield (t·ha^{-1}) of forage rapeseed at four growth stages (bud period, initial flower-

ing, final flowering, podding) as affected by transplanting date. The experimental treatments were five transplant dates (month-day): 10-20, 10-30, 11-10, 11-20, and 11-30. Different lowercase letters indicate significant differences among different transplanting dates in each growth stage at the 0.05 level. ns, no significant differences among different transplanting dates.

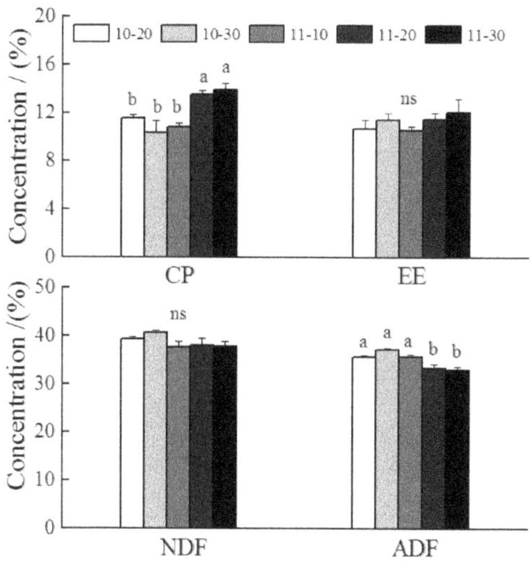

Figure 2. Nutritional content (%) of forage rapeseed at podding as affected by transplanting date. The experimental treatments were five transplant dates (month-day): 10-20, 10-30, 11-10, 11-20, and 11-30. The four measures of nutritional quality were crude protein (CP), ether extract (EE), neutral detergent fiber (NDF), and acid detergent fiber (ADF). Different lowercase letters indicate significant differences among different transplanting dates at the 0.05 level. ns, no significant differences among different transplanting dates.

3.2. *Effects of Transplant Date on Grain Yield and Yield Components of Rapeseed*

The rapeseed grain yield was significantly related to the transplanting date. The highest yield was 3922 kg·ha^{-1} in the transplanting treatment of 20 October, which was not significantly different from that in the transplanting treatment of 30 October. However, the yields in the transplanting treatments in November were significantly lower than that in the 20 October transplanting treatment by 17.2% to 22.5% (Figure 3).

Compared with the 20 October transplanting treatment, the number of primary branches decreased significantly by 31.2% in the 30 November transplanting treatment, while the number of secondary branches decreased significantly by 40.8% in the 10 November, 78.2% in the 20 November, and 57.1% in the 30 November transplanting treatments (Table 2). The numbers of primary and secondary branches were similar in the 20 and 30 October transplanting treatments (Table 2). With the extension in transplanting date, the numbers of pods in the main inflorescence and the primary branches firstly increased and then decreased (Table 2). However, compared with the earliest transplanting treatment of 20 October, in the other treatments, the number of pods on secondary branches decreased significantly by 26.9% (30 October), 69.7% (10 November), 82.4% (20 November), and 78.4% (30 November) (Table 2). There were no significant differences in the number of seeds in the main inflorescence among treatments. The number of seeds on the branches was lower in the 20 and 30 November transplanting treatments than in that of 20 October. The thousand-seed weight was not significantly different among transplanting date treatments (Table 2). Rapeseed grain yield was significantly positively correlated with the number of secondary branches, pod number of secondary branches, and seed number of secondary branches (Table 3).

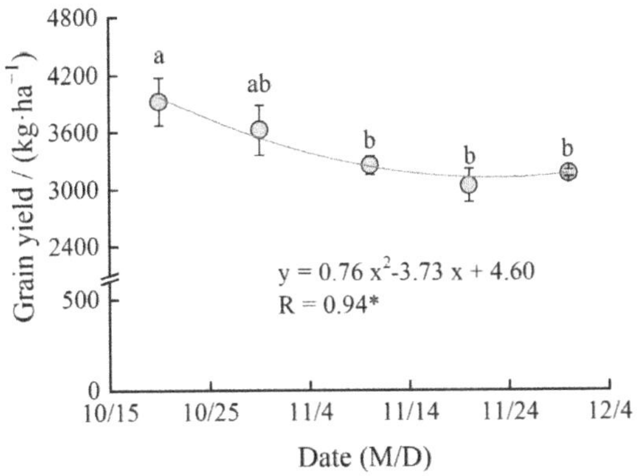

Figure 3. Grain yield (kg·ha^{-1}) of rapeseed at maturity as affected by transplanting date (month/day) and the *Pearson* coefficients. The experimental treatments were five transplant dates (month-day): 10-20, 10-30, 11-10, 11-20, and 11-30. Different lowercase letters indicate significant differences among different transplanting dates at the 0.05 level. * $p < 0.05$.

Table 2. Yield components of rapeseed at maturity as affected by transplanting date.

Transplanting Date (Month-Day)	Branch Number per Plant		Pod Number per Plant			Seed Number per Pod per Plant			1000 Seed Weight (g)
	Primary Branches	Secondary Branches	Main Inflorescence	Primary Branches	Secondary Branches	Main Inflorescence	Primary	Secondary	
10-20	7.7 a	14.7 a	73.7 b	306.0 ab	237.7 a	25.0 a	24.6 ab	22.9 a	3.9 a
10-30	8.3 a	13.0 a	79.0 ab	352.0 a	173.7 b	23.9 a	21.9 ab	24.2 a	4.1 a
11-10	9.0 a	8.7 b	92.0 a	367.7 a	72.0 c	24.6 a	25.5 a	25.1 a	4.1 a
11-20	7.3 ab	3.2 c	86.0 ab	254.2 bc	41.8 c	21.9 a	22.5 ab	17.2 b	3.9 a
11-30	5.3 b	6.3 b	70.7 b	211.3 c	51.3 c	21.6 a	20.6 b	19.6 ab	4.1 a

Values for transplanting dates with different lowercase letters representing significantly different at the same ensiling days at the 0.05 level.

Table 3. Pearson coefficients of correlation between grain yield and yield components at maturity.

Yield Component	*r* Coefficient	Pr > F
Primary branch number	0.333	ns
Secondary branch number	0.795	***
Pod number of main inflorescence	−0.072	ns
Pod number of primary branches	0.448	ns
Pod number of secondary branches	0.834	***
Seed number of main inflorescence	0.272	ns
Seed number of primary branches	−0.031	ns
Seed number of secondary branches	0.670	**
1000 seed weight	0.186	ns

** $p < 0.01$; *** $p < 0.001$; ns, not significant.

3.3. *Effects of Transplant Date on Growth Characteristics of Rapeseed*

With the RCPPS, all treatments had comparable leaf age at the seedling stage. However, compared with the 20 and 30 October transplanting treatments, the transplanting treatments in November had significantly lower leaf age in the late growth period (Figure 4). From budding to flowering stages, late transplanting significantly inhibited the rhizome thickness, whereas at the podding stage, no significant differences were found among transplanting dates (Figure 5). The transplanting date had significant effects on plant height from flowering to podding stages (Figure 6). With the delay in transplanting date, the plant height tended to increase and then decrease. At the podding stage, the lowest plant height was in the 30 November transplant treatment (Figure 6).

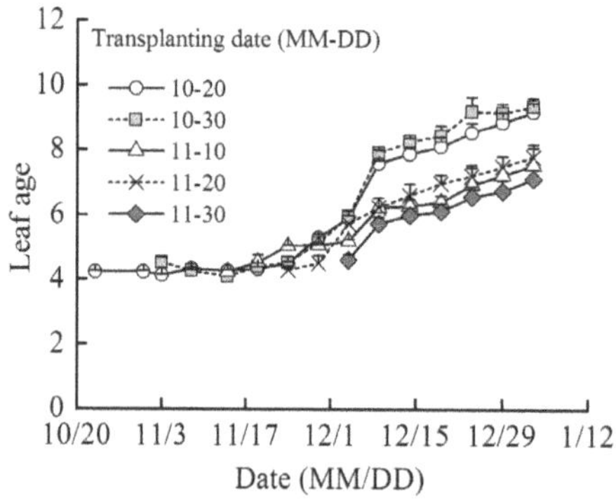

Figure 4. Dynamics of rapeseed leaf age as affected by transplanting date (month/day). The experimental treatments were five transplant dates (month-day): 10-20, 10-30, 11-10, 11-20, and 11-30.

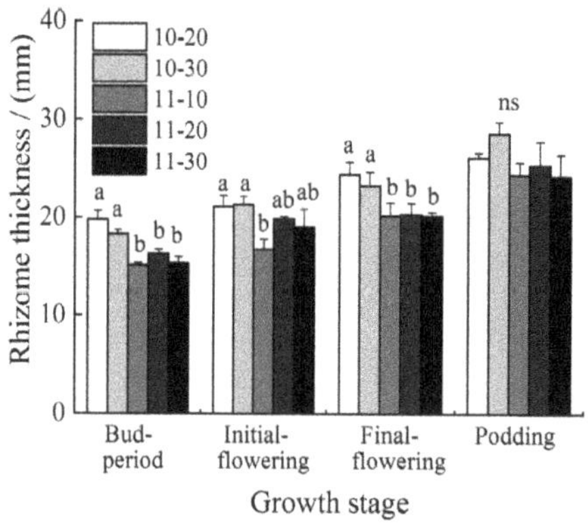

Figure 5. Rhizome thickness (mm) of forage rapeseed at four growth stages (bud period, initial flowering, final flowering, podding) as affected by transplanting date. The experimental treatments were five transplant dates (month-day): 10-20, 10-30, 11-10, 11-20, and 11-30. Different lowercase letters indicate significant differences among different transplanting dates in each growth stage at the 0.05 level. ns, no significant differences among different transplanting dates.

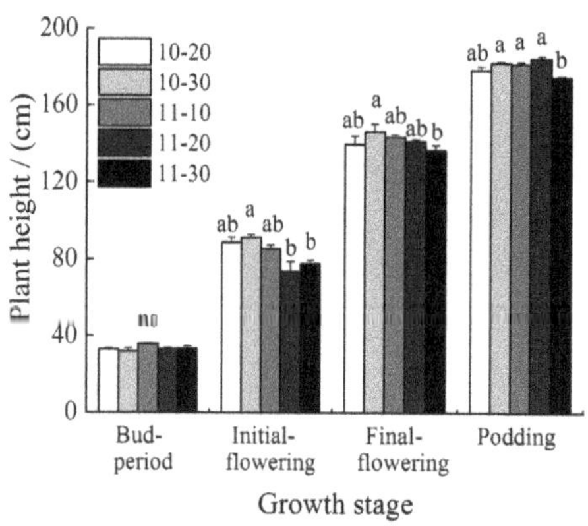

Figure 6. Plant height (cm) of forage rapeseed at four growth stages (bud period, initial flowering, final flowering, podding) as affected by transplanting date. The experimental treatments were five

transplant dates (month-day): 10-20, 10-30, 11-10, 11-20, and 11-30. Different lowercase letters indicate significant differences among different transplanting dates in each growth stage at the 0.05 level. ns, no significant differences among different transplanting dates.

4. Discussion

Dry matter yield is one of the important factors used to evaluate the productivity of forage [26,27]. In this study, the biomass yield of forage rape increased rapidly from budding to final flowering stage, followed by a gradual increase in the podding stage (Figure 1). Regardless of the transplanting date, the highest dry matter yield was in the podding stage, reaching 6.4 $t{\cdot}ha^{-1}$ in the 30 October transplant date treatment. This forage yield accumulation is consistent with that of the rape blanket seedlings transplanted technique, which has a maximum forage yield of approximately 6.3 $t{\cdot}ha^{-1}$ in the Yangtze River delta [28]. The growth rate was highest at the bud period and then decreased after flowering. However, the yield of forage rape in Gansu reaches 11 $t{\cdot}ha^{-1}$, which is approximately two times higher than the yield in the Yangtze River delta [12]. The difference in yield is likely primarily due to different experimental sites with specific soil characteristics and different varieties and planting methods. For example, the rapeseed in Northwest China is primarily directly sowed mechanically after the harvest of wheat, and the crops show vigorous growth, particularly in the early growth period. In addition, in the northern region, a waterlogging disaster during the growing season is unlikely. However, in this study, low temperatures and continuous rainfall occurred during the transplanting and the seedling growth period, with the cumulative precipitation reaching 414.7 mm in October and November (Table 1), which may also affect the soil nutrient availability, severely inhibiting the growth and yield formation of rapeseed before winter. The nutritional value of forage grass primarily depends on the crude protein content, whereas the digestibility of forage primarily depends on the ADF content [12]. Of the main nutrient measures of forage rape at the podding stage (Figure 2), the crude protein content was 11.6% to 14.0%, the NDF content was 37.7% to 40.7%, and the ADF content was 33.0% to 37.2%, which meet the classification standards for legumes in the United States [29].

At the final flowering stage, the dry matter yield of forage rape was higher in the October transplant date treatments than in the November transplant date treatments, whereas there was no significant difference among transplant date treatments at the podding stage (Figure 1). Relatively few studies have examined the effect of transplanting date on the characteristics of forage rape with RCPPS. With the rape blanket seedling transplant technique, the biomass yield of forage rape transplanted before 20 October is significantly higher than that of rape transplanted in late October in the Yangtze River delta [28]. With the direct-seeding technique, forage rape should be sowed earlier, because the accumulation of dry matter yield is primarily limited by air temperature, particularly in northern China. The temperature always declines rapidly in August in Gansu Province, which significantly inhibits the dry matter formation of rape in the late growth period. From 30 July to 8 August, the biomass yield of rape was significantly decreased by 50% due to the delay of 5 days [10]. In addition, the yield and plant height of forage rape were both higher when sowed in mid-July than in early August in Heilongjiang Province [14]. Notably, with the delay in transplanting in the 20 and 30 November treatments, the crude protein content increased significantly, whereas the ADF content decreased significantly (Figure 2). By contrast, the increase in ADF would reduce the amount of digestible dry matter, influencing the digestibility of forage [29]. The duration of the growing season is decided by the sowing date, which therefore influences the accumulation of biomass yield and nutrients [30]. In Heilongjiang Province, the contents of crude protein, crude fat, carbohydrates, and trace elements in early sown forage rape were all higher than those in late-sown rape [12]. Overall, we conclude that transplanting in late November with RCPPS for forage use has distinct advantages in the rice–rape rotation, as it can maintain yield and improve nutritional value of forage rape, so as to alleviate the stubble contradiction between late harvested rice in the Yangtze River delta. All these findings can substantially improve the annual economic benefits for growing rice–rape crops, which will promote the application of the rice-rape rotation system.

Considering the conventional use of rapeseed for grains, the rapeseed transplanted on 20 and 30 October had comparable grain yields. However, compared with the 20 October transplant date, the rapeseed transplanted in November had significantly lower grain yields by 17.2% to 22.5% (Figure 3). Grain yield is determined by the number of pods per plant, the number of seeds per pod, and the weight of one thousand seeds. Moreover, the number of pods per plant is considered to be the most important factor in determining yield, which is also easily influenced by agricultural management [31]. Compared with traditional transplanting methods, the increase in yields in the RCPPS is primarily attributed to the number of pods [17]. With the RCPPS in this study, the grain yield was most affected by the number of secondary branches, as indicated by the significant positive correlations between yield and the number of secondary branches and the number of pods and seeds on secondary branches (Table 3). Compared with the earliest transplanting treatment, the number of pods on secondary branches in the November transplanting treatments decreased significantly by 69.7% to 82.4% (Table 2). The late transplanting dates significantly inhibited leaf growing stage and resulted in a reduced number of effective leaves and diameter of pre-winter root (Figures 4 and 5, respectively). Xiong et al. (2016) [15] found that substantial dry matter accumulation and the number of effective pods promote high grain yield under the appropriate planting date. In Jiangxi Province, with a simplified transplanting technique, the yield was optimum when sown on 20 September and then nursed for nearly 40 days. Changes in the quantitative processes of yield under different planting dates with RCPPS should be explained in terms root morphology and physiology in future.

5. Conclusions

In conclusion, in the rice–rape rotation system with the RCPPS in the Yangtze River delta, transplanting rape from 20 to 30 October led to higher grain yields by 19.6% on average, which were significantly correlated with numbers of secondary branches, pods, and seeds. Thus, relatively early maturing rice varieties should precede the rape crop. Alternatively, to alleviate the competition for resources between rape and rice crop seasons, forage rape can be transplanted in mid-November and harvested from the final flowering to podding stages to improve forage yield and nutritional value, particularly with improved crude protein content and reduced acid detergent fiber content. Thus, considering the security of oil and forage production, the right transplanting time of specific use could be adopted for farmers in RCPPS to improve the value of rapeseed meanwhile ensure the yield of previous rice crops, which is flexible and sustainable in multi-cropping rice rotation system in Yangtze River delta.

Author Contributions: Conceptualization, Y.T.; Investigation, Y.T., C.L., M.H. and H.W.; Data curation, Y.T., D.L. and Y.Y.; Writing—original draft, Y.T.; Writing—review & editing, H.S.; Supervision, H.S.; Funding acquisition, H.S. All authors have read and agreed to the published version of the manuscript.

Funding: This research were funded by [National Natural Science Foundation of China] grant number [32101854] and [China Agriculture Research System of MOF and MARA] grant number [CARS-12]

Institutional Review Board Statement: Not applicable.

Informed Consent Statement: Not applicable.

Data Availability Statement: The datasets used and analyzed during the current study are available from the corresponding author upon reasonable request.

Acknowledgments: We are grateful to the researcher Mingxing Shen of the Jiangsu Suzhou Executive Leadership College for guidance in the experiment design, as well as the technical and laboratory assistants of Yangzhou University and Ludong University, Yunlong Tang, Yuquan Mao, Qingsong Zhang, and Kaiyue Yuan, for their help in the field and laboratory work.

Conflicts of Interest: We declare that we have no financial and personal relationships with other people or organizations that could inappropriately influence our work. There is no professional or other personal interest of any nature or kind in any product, service, or company that could be construed as influencing the position presented in, or the review of, the manuscript.

Appendix A

Figure A1. Ridge culture place planting system (RCPPS) for rapeseed (Suzhou Comprehensive Experimental Station of China Rape Research System).

References

1. Fang, Y.T.; Ren, T.; Zhang, S.T.; Liu, Y.; Liao, S.P.; Li, X.K.; Cong, R.H.; Lu, J.W. Rotation with oilseed rape as the winter crop enhances rice yield and improves soil indigenous nutrient supply. *Soil Tillage Res.* **2021**, *212*, 105065. [CrossRef]
2. Guan, C.Y.; Jin, F.R.; Dong, G.Y.; Guan, M.; Tan, T.L. Exploring the growth and development properties of early variety of winter rapeseed. *Eng. Sci.* **2012**, *14*, 4–12.
3. Jiang, F.Y.; Tang, S.; Wang, J.J.; Zhang, D.X.; Tian, X.C.; Chen, A.W.; Zhou, G.S.; Wu, J.S. Technique of direct sowing rapeseed in no-tillage paddy rice field. *Hubei Agric. Sci.* **2010**, *49*, 2372–2375.
4. Wang, Y.; Wang, Y.; Lu, J.W.; Li, X.K.; Ren, T.; Cong, R.H. Response differences in growth and yield formation of direct-sown and transplanted winter oilseed rape to N, P and K fertilization. *J. Plant Nutr. Fertil.* **2016**, *22*, 132–142.
5. Shen, J.X.; Fu, T.D. Rapeseed production, improvement and edible oil supply in China. *J. Agric. Sci. Technol.* **2011**, *13*, 1–8.
6. Li, Y.L.; Liu, A.M. Competition mechanism of cultivated land resources in winter agriculture in the Yangtze basin: Theory and case study. Resour. Environ. *Yangtze Basin* **2009**, *18*, 146–151.
7. Islam, M.R.; Garcia, S.C.; Horadagoda, A. Effects of residual nitrogen, nitrogen fertilizer, sowing date and Harvest time on yield and nutritive value of forage rape. *Anim. Feed. Sci. Technol.* **2012**, *177*, 52–64. [CrossRef]
8. Kaur, R.; Garcia, S.C.; Fulkerson, W.J. Feeding time and sequence of forage rape and maize silage does not affect digestibility and rumen parameters in sheep. *Anim. Prod. Sci.* **2009**, *49*, 318–325. [CrossRef]
9. Liu, Y.S.; Chen, Z.F.; Feng, W.L.; Cao, Z. The planting technology and industrial development prospects of forage rape in the Loess Hilly Area-A case study of newly-increased cultivated land through Gully Land Consolidation in Yan'an, Shaanxi Province. *J. Nat. Resour.* **2017**, *32*, 2065–2074.
10. Fu, T.D.; Tu, J.X.; Zhang, Y.; Chen, C.B.; Yang, Q.F.; Teng, H.Y.; Niu, J.L.; Yin, J.Z. Study and utilization of the forage rapeseed after harvesting wheat in the northwest of China. *Sci. Technol. West China* **2004**, *6*, 4–7.

11. Wang, H.C.; Liu, D.S.; Liu, C.L.; Lu, X.C.; Wei, W.; Wang, X.X. Research progress of forage rape and its feed value in China. *Soils Crops* **2016**, *5*, 60–64.
12. Yang, R.J. Effect of planting density on growth properties of rapeseed in wheat/silage rape multiple cropping. *Chin. J. Oil Crop Sci.* **2007**, *29*, 479–482.
13. National Agricultural Technology Extension Service Center, Ministry of Agricultural and Rural Affairs of the People's Republic of China. Winter Rapeseed Production Technical Guidance Opinions 2017–2018. 2017. Available online: http://www.moa.gov.cn/gk/nszd_1/2017nszd/201709/t20170927_5828755.htm (accessed on 27 September 2017).
14. Liu, M.; Xiao, J.L.; Li, W.; Bi, Y.D.; Di, S.F.; Liu, M.; Lai, Y.C. Effects of different sowing date on the yield and quality indices of multi-cropping forage rape after reaping wheat in Northern Alpine Regions. *J. Anhui Agric. Sci.* **2014**, *42*, 12933–12934.
15. Xiong, J.; Ding, G.; Dai, X.Y.; Chen, L.L.; Li, S.Y.; Zou, X.F.; Zou, X.Y.; Song, L.Q. Effects of sowing date and seedling age on growth., development and yield of simply-raised transplanted rapeseed. *Acta Agric. Jiangxi* **2016**, *28*, 14–17.
16. Zhang, C.N.; Luo, T.; Liu, J.H.; Xian, M.Z.; Yuan, J.Z.; Hu, L.Y.; Xu, Z.H. Evaluation of the low temperature tolerance of rapeseed genotypes at the germination and seedling emergence stages. *Crop Sci.* **2019**, *59*, 1709–1717. [CrossRef]
17. Sun, H.; Huang, M.; Chen, P.F.; Zhang, J.D.; Lu, C.Y.; Qiao, Z.Y.; Song, Y.; Wang, H.H.; Sheng, M.X. Effects of rice straw returning and transplanting methods on rape yield and economic benefit. *Chin. Agric. Sci. Bull.* **2015**, *31*, 121–125.
18. Huang, M.; Zhang, W.X.; Jiang, L.G.; Zou, Y.B. Impact of temperature changes on early-rice productivity in a subtropical environment of China. *Field Crops Res.* **2013**, *146*, 10–15. [CrossRef]
19. Keeney, D.R. Nitrogen-availability indices. In *Methods of Soil Analysis: Part 2. Chemical and Microbiological Properties*; Page, A.L., Miller, R.H., Keeney, D.R., Eds.; SSSA Publications, Inc.: Madison, WI, USA, 1982; pp. 711–730.
20. Nelson, D.W.; Sommers, L.E. *Methods of Soil Analysis Part II. Chemical and Microbiological Properties*; American Society of Agronomy, Soil Science Society of America: Madison, WI, USA, 1982.
21. Olsen, S.R.; Cole, C.V.; Watanabe, F.S.; Dean, L.A. *Estimation of Available Phosphorous in Soil by Extracting with Sodium Bicarbonate*; USDA Circular 939: Washington, DC, USA, 1954.
22. Bao, S.D. *Agrochemical Analysis of Soil*; China Agricultural University Press: Beijing, China, 2002; pp. 25–38.
23. Van Soest, P.J.; Sniffen, C.J.; Mertens, D.R. A net protein system for cattle: The rumen sub-model for nitrogen. In Proceedings of the International Symposium on Protein Requirements for Cattle, Oklahoma City, OK, USA, 21–19 November 1981; Owens, F.N., Ed.; Oklahoma State University: Stillwater, OH, USA, 1981; p. 265.
24. Zhang, L.Y. *Feed Analysis and Feed Quality Inspection Technology*, 3rd ed.; China Agricultural University Press: Beijing, China, 2007.
25. SAS Institute. *SAS User's Guide: Statistics, Version 8.2*, 6th ed.; SAS Institute: Cary, NC, USA, 2009.
26. Jia, C.H.; Qian, W.X.; Tursunay, S.; Ao, W.P.; Abudukeranmu, G. Roughage nutritional value evaluation indices and research methods. *Prataculturai. Sci.* **2017**, *34*, 415–427.
27. Tao, Y.Y.; Shen, X.W.; Xu, J.; Shen, Y.; Wang, H.H.; Lu, C.Y.; Shen, M.X. Characteristics of heat and solar resources allocation and utilization in rice-oilseed rape double cropping systems in the Yangtze River Delta. *Acta Agron. Sin.* **2023**, *49*, 1326–1337.
28. Tao, Y.Y.; Tang, Y.L.; Xu, J.; Wang, H.H.; Huang, M.; Sun, H.; Shen, M.X. The forage rapeseed yield and nutrition on carpet-seedling transplantation at different planting date. *Pratacultural Sci.* **2019**, *36*, 1–8.
29. Cheng, L.X.; Yang, R.J.; Ge, G.T.; Sun, L.; Fu, J.P.; Jia, Y.S. Measurement and evaluation of grading indexes (GI) of five forages. *Pratacultural Sci.* **2013**, *30*, 1284–1288.
30. Liu, Z.Q. *Study on Sowing Date Effect on Development of Rapeseed*; Huazhong Agricultural University: Wuhan, China, 2008.
31. Diepenbrock, W. Yield analysis of winter oilseed rape (*Brassica napus* L.): A review. *Field Crops Res.* **2000**, *67*, 35–49. [CrossRef]

applied sciences

MDPI

Article

Correlation between Spring Wheat Physiological Indicators and UAV Digital Image Index in Hetao Irrigation Area

Min Xie [1], Jun Luo [1,2], Lijun Li [2], Peng Zhang [1], Qiang Wu [1,3], Mengyuan Li [1], Haixia Wang [2] and Yongping Zhang [1,*]

[1] College of Agriculture, Inner Mongolia Agricultural University, Hohhot 010010, China; xiemin@imau.edu.cn (M.X.); imauwq@163.com (Q.W.)
[2] Inner Mongolia Technology Extension Center of Agriculture and Animal Husbandry, Hohhot 011799, China
[3] Inner Mongolia Academy of Agriculture and Animal Husbandry, Hohhot 010031, China
* Correspondence: imauzyp@163.com

Abstract: To accurately and non-destructively monitor the growth of spring wheat in the Hetao irrigation area, UAV remote sensing was employed during various fertility stages. Digital image indices from diverse fertilization treatments were calculated and compared with physiological indices to identify the most sensitive digital image indices corresponding to these indices. The study underscored the critical importance of the flowering stage in the growth of spring wheat, thus highlighting the necessity of focusing on this stage. This finding reiterated that the flowering stage was pivotal for spring wheat development in the Hetao Irrigation Area. Several digital image indices, such as GLA, R, G, INT, g, GRVI, MGRVI, RGBVI, EXG, and GRRI, exhibited a high frequency of significant correlations with physiological indices during the four primary reproductive stages of wheat. Consequently, these sensitive digital image indices during the flowering stage can more effectively characterize the physiological indices of spring wheat.

Keywords: spring wheat; UAV; digital image index physiological indicators

1. Introduction

Remote sensing technology has advanced agricultural quantitative remote sensing, making it the predominant and precise method for assessing global plant growth. A mathematical model established relationships between image information derived from remote sensing sensors and specific plant parameters. Analyzing image characteristics facilitated the monitoring, diagnosis, and comprehension of plant growth status, enabling the accurate regulation and management of plant growth [1]. Several studies showed that crop quantitative remote sensing was effective in reflecting actual nitrogen levels and monitoring nitrogen across various cereal crops [2–4]. Mullan [5] used a digital image analysis tool for high-throughput screening of four large wheat populations to validate the relationship between digital image analysis and measures of the normalized difference vegetation index (NDVI), leaf area index, and light penetration through the crop canopy. Image processing techniques have immense potential to use characteristics like size, shape, color, and texture attributes from digital images to characterize agricultural produce [6]. Due to the ability of crop stems, leaves, and other aboveground organs to absorb, reflect, and refract visible light, with advantages of mobility, timeliness, ease of operation, and high resolution, light parameters such as color, hue, brightness, and saturation could enable a comprehensive characterization of plants, and digital image index made it possible to standardize spectral information from visible light bands, specifically red (R), green (G), and blue (B) [7,8].

Unmanned aerial vehicles (UAVs) for remote sensing, with their advantages of mobility, timeliness, ease of operation, and high resolution, have recently shown significant advantages in extracting crop information and monitoring nutrient levels across small to medium spatial scales [9–11]. The traditional determination of crop physiological indicators and nutrient diagnosis required a large number of field samples and laboratory analysis, which was cumbersome and time-consuming. With the development of digital image technology, some researchers began to apply digital image processing technology for nutrient diagnosis [12]. Xia et al. [13] used a cell phone camera to obtain the canopy images of maize in different periods, extracted digital image indexes from the images, and analyzed the correlation between the digital image indexes and maize nitrogen nutrient indexes. Their experimental results showed that, in the six-leaf period of maize, the color parameter of the canopy image was significantly correlated with the maize SPAD value, indicating it could be used as a nitrogen diagnostic period. Moreover, B/(R + G + B) and G/(R + G + B) were significantly correlated with the leaf SPAD value, a traditional indicator of nitrogen diagnosis, and B/(R + G + B) was more sensitive, so it could be used as a color parameter indicator for the diagnosis of nitrogen nutrition in maize. This result suggested that it was feasible to use digital image technology for nitrogen nutrition diagnosis. In addition, Wang et al. [14] used a digital camera to obtain the image data of rice canopy, extracted the image index and the actual nutrient index of rice for correlation analysis, and found that the two had a good correlation relationship, in which the correlation coefficient between NRI and leaf nitrogen content reached −0.65, indicating that the application of image technology could be fast and simple to realize the rapid monitoring of crop nutrient indexes.

The Hetao region of Inner Mongolia is a significant center for spring wheat production in China. Research into the rapid, non-destructive, and cost-effective monitoring of spring wheat production using unmanned aerial vehicles (UAVs) is indispensable for this region's agricultural sector. This study conducted field trials on spring wheat with varying nitrogen levels and captured digital images using UAVs at different reproductive stages. Plant morphological and physiological yield indicators were also collected and analyzed. Eighteen digital image indices were selected for analysis, and correlation analyses were used to determine the most effective diagnostic digital image indices across various stages.

2. Materials and Methods

This study conducted simultaneous field wheat fertilizer control trials at two locations: Xinhua Agricultural Cooperative Farm, Linhe district, and Yonglian village, Xinguangzhong town, Wuyuan county Bayannur city, Inner Mongolia Autonomous Region in 2019 and 2020. The locations of the experimental plots are shown in Figure 1.

The test variety was the locally dominant wheat variety, Yongliang 4. Sowing took place in early March 2019 and 2020, and harvesting took place in late July 2019 and 2020. The seeding rate was 375 kg/hm^2, using a wheat planter for seed fertilizer layer seeding with a row spacing of 15 cm. Nitrogen, phosphorus, and potash fertilizers were applied together as the seed fertilizer at the time of sowing, followed by an additional treatment during the jointing stage (3) of the wheat. Irrigation was carried out in a water-saving mode, twice during the whole lifecycle (jointing (3) + flowering (6), each irrigation 900 m^3/hm^2). Other management measures were in line with conventional wheat cultivation practices. The fertilization experiment included six different modes: CK, farmer model (FM), control of fertilization patterns (CF), control of fertilization-N (CF-N0), control of fertilization-P (CF-P0), and control of fertilization-K (CF-K0). Details of the specific fertilization rates and methods are shown in Table 1.

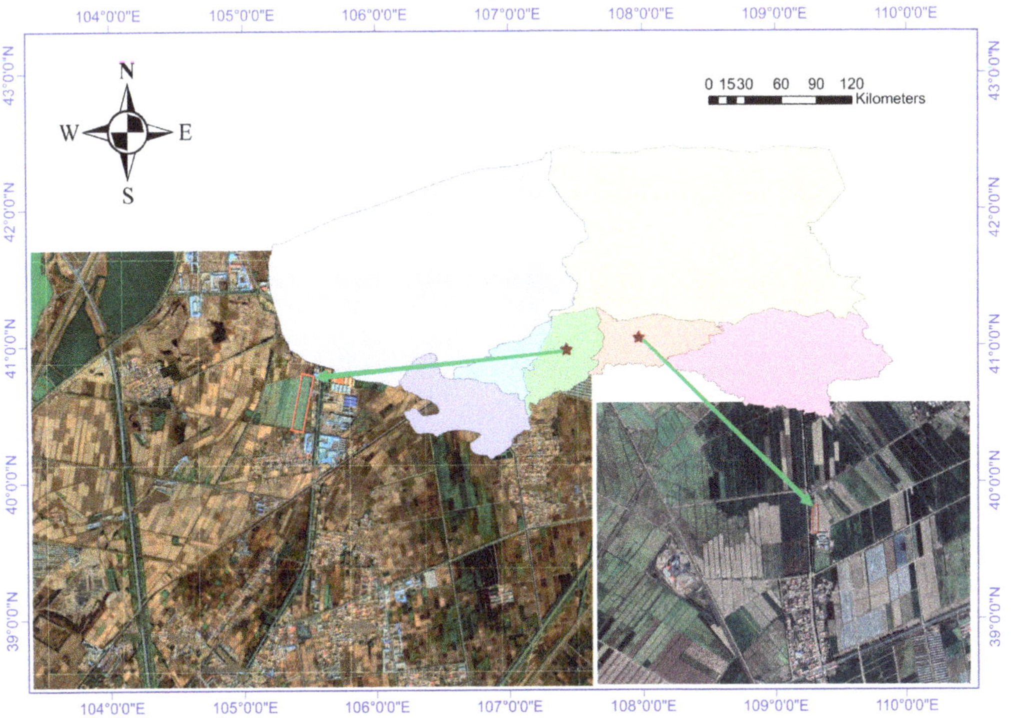

Figure 1. Map of test site locations.

Table 1. This table displays the fertilization rates of treatments.

Treatments	Fertilizer Rate (kg/hm^2)			
	Nitrogen		Phosphate Fertilizer (P_2O_5)	Potash Fertilizer (K_2O)
	Sowing N	Top-Dressing N		
CK	0	0	0	0
FM	94.5	207	241.5	0
CF	52.5	124.5	61.5	60
CF-N0	0.0	0.0	61.5	60
CF-P0	52.5	124.5	0.0	60
CF-K0	52.5	124.5	61.5	0.0

Digital image data were collected in late May (tasseling stage (5)), mid-June (filling stage (7)), and early July (maturity stage (8)) in 2019. A DJI Royal MAVIC 2 Professional Edition (Manufactured by DJI Innovations (Shenzhen, China) Co., Ltd., DJI unmanned aerial vehicles (UAVs) are produced at the DJI UAV production base in Chengdu, China. All lenses were purchased from the same company.) equipped with a Hasselblad L1D-20c camera was used for high-definition digital photography. The camera features a new 1-inch 20-megapixel CMOS sensor, a built-in wide-angle lens with a 28 mm equivalent focal length, an adjustable aperture value ranging from f/2.8 to f/11, and a maximum sensitivity of 12,800. The study collected and stored digital images in the TIF format. The drone's flight lasted approximately 30 min. To minimize the impact of external factors on image acquisition, the study was conducted between 10:00 and 14:00 under clear and cloudless sky conditions. Radiometric correction was carried out using a black-and-white board to minimize errors, and the UAV was flown at an altitude of 50 m.

Digital image indices have been effectively used to estimate plant physiological and biochemical parameters, such as leaf area index, biomass, and nitrogen content. This effectiveness was due to their ability to enhance vegetation information, reduce the impact of non-vegetation elements on plant mapping features, and improve the accuracy of remote sensing inversion. This study screened digital image indices for monitoring wheat growth research, integrating insights from previous studies with a pre-analysis of experimental data. The name of each digital image index and the formula used to calculate it are shown in Table 2 below:

Table 2. Calculation of digital image indices and cited literature sources. R, G, B, and near infrared (NIR) indicate the measured reflectance (digital numbers) in red, green, blue, and near-infrared wavelength ranges, respectively.

Digital Image Index	Description	Calculating Formula	Source Literature Reference
R	Red	R	-
G	Green	G	-
B	Blue	B	-
r	Normalized Red Index	R/(R + G + B)	[15]
g	Normalized Green Index	G/(R + G + B)	[15]
b	Normalized Blue Index	B/(R + G + B)	[15]
ExG	Excess Green	2 × G − R − B	[8]
GRRI	Green–Red Ratio Index	G/R	[8]
GBRI	Green–Blue Ratio Index	G/B	[8]
RBRI	Red–Blue Ratio Index	R/B	[8]
GRVI	Green–Red Vegetation Index	(G − R)/(G + R)	[8]
INT	Intensity	(R + G + B)/3	[15]
IKAW	Normalized Red–Blue Index	(R − B)/(R + B)	[8]
MGRVI	Modified Green–Red Vegetation Index	$(G^2 - R^2)/(G^2 + R^2)$	[8]
RGBVI	Red–Green–Blue Vegetation Index	$(G^2 - B \times R)/(G^2 + B \times R)$	[8]
GLA	Green Leaf Area	(2 × G − R − B)/(2 × G + R + B)	[8]
CIVE	Color Index of Vegetation Extraction	0.441R − 0.881G + 0.385B + 18.7875	[8]

The study examined plant parameters, including leaf area (PLA), leaf area index (LAI), chlorophyll SPAD, dry matter accumulation (DMA), fluorescence parameters (F_v/F_m), intercellular CO_2, concentration (C_i), stomatal conductance (Gs), net photosynthetic rate (NPR), transpiration rate (T_r/E), water use efficiency (WUE), spike number (PN), grain number (GN), thousand grain weight (TKW), yield (Y), and plant nitrogen content (NC).

The study also measured the stem and leaf nitrogen content (SLNC), stem and leaf phosphorus content (SLPC), stem and leaf potassium content (SLKC), seed nitrogen content (KNC), seed phosphorus content (KPC), kernel potassium content (KKC), glume nitrogen content (GNC), phosphorus content of glumes (GPC), potassium content of glumes (GKC), number of spikes (PN), number of grains in spikes (GN), thousand kernel weight (TKW), and yield (Y), with corresponding vegetation indices.

The study utilized the correlation coefficient (r) to measure the degree of correlation between wheat digital image indices and physiological growth characterization parameters. The coefficient was calculated using the following formula:

$$r = \frac{\sum (x - \overline{x})(y - \overline{y})}{\sqrt{\sum (x - \overline{x})^2 \cdot \sum (y - \overline{y})^2}} \quad (1)$$

The correlation coefficient (r) indicates the degree of linear relationship between two variables. A correlation between two variables can be positive, negative, or non-existent, indicated by $1 \geq r > 0$, $-1 \leq r < 0$, and r = 0, respectively. The absolute value of r indicates the strength of the correlation. To determine the significance of the correlation coefficient, refer to the critical value table for correlation coefficients. Significance levels are denoted by * and ** at the 0.05 and 0.01 levels, respectively.

As indicated in Table 3, the analysis of variance (ANOVA) demonstrated significant differential effects of fertilization treatments on the digital image indices G, B, and INT. No significant effects were observed on the remaining indices.

Table 3. Analysis of variance (ANOVA) of digital image indices under different fertilization treatments. (Bolded markers are parameters that were significant or highly significant under different treatments. Then, after ANOVA analysis, 'a' represents the value with the highest mean at a significant level, 'b' represents values significantly different from it, and so forth).

Digital Image Indices	*p*-Value	CK	FM	CF	CF-N0	CF-P0	CF-K0
R	0.058	172.68	136.87	135.57	158.85	117.82	113.33
G	**0.007**	**171.64 a**	**153.92 abc**	**149.85 bc**	**156.93 ab**	**136.97 c**	**136.00 c**
B	**0.011**	**109.60 a**	**97.31 ab**	**92.59 abc**	**93.67 abc**	**76.49 bc**	**73.69 c**
EXG	0.580	64.11	22.52	28.71	67.09	22.18	16.96
INT	**0.005**	**151.30 a**	**129.37 abc**	**126.00 bc**	**136.48 ab**	**110.42 c**	**107.67 c**
CIVE	0.906	−14.08	−18.99	−17.80	−13.36	−20.49	−22.68
r	0.791	0.38	0.35	0.36	0.39	0.35	0.35
g	0.414	0.38	0.40	0.40	0.39	0.41	0.42
b	0.674	0.24	0.25	0.25	0.23	0.23	0.23
GRRI	0.634	1.01	1.17	1.15	1.00	1.20	1.24
GBRI	0.181	1.59	1.59	1.63	1.70	1.80	1.86
RBRI	0.859	1.58	1.41	1.47	1.71	1.56	1.58
GRVI	0.656	0.00	0.07	0.06	0.00	0.08	0.10
IKAW	0.844	0.22	0.16	0.18	0.26	0.21	0.21
MGRVI	0.666	0.00	0.13	0.11	−0.00	0.16	0.18
RGBVI	0.318	0.22	0.29	0.29	0.25	0.36	0.38
GLA	0.419	0.10	0.14	0.14	0.11	0.17	0.18

The data of wheat physiological indexes in this experiment were determined by a portable photosynthesis meter (Manufacturer: Hansatech Instruments Ltd.—China Headquarters of PP Systems (Affiliated with Hansatech Instruments Ltd.), Amesbury, MA, USA), chlorophyll fluorometer (The SPAD-502PLUS chlorophyll meter is manufactured by Konica Minolta Sensing, Inc. in Osaka, Japan), Kjeldahl nitrogen meter (The fully automatic Kjeldahl nitrogen analyzer (K9860) is manufactured by Shanghai Licheng Scientific Instrument Co., Ltd., Shanghai, China), and FOSS grain analyzer (The FOSS Infratec TM 1241 near-infrared grain quality analyzer is manufactured by Foss (Beijing, China) Science and Trade Co., Ltd.). Adobe Photoshop CC 2018 software was used for digital image processing. Non-wheat leaf information was excluded from the experimental plots after segmenting the images based on different fertilization treatments. The red (R), green (G), and blue (B) values from each treatment plot and the two-year trial plots were extracted and used to compute the average of each digital image index. The R programming language was used for calculating, correlating, analyzing, and graphing each variable.

3. Results

3.1. Relationships between Digital Image Indices and Wheat Physiological Indicators at the Time of Jointing and Flowering

The analysis involved correlating wheat physiological indices SPAD, LAI, LAD, and DMA with each digital image index. During the wheat jointing stage (3), the SPAD value showed a highly significant negative correlation with the g, RGBVI, and GLA indices. Additionally, it showed a significant correlation with the GRRI, GBRI, GRVI, and MGRVI indices and a significant positive correlation with the R, G, B, INT, and CIVE indices. The LAI values showed a significant negative correlation only with R. The LAD had a highly significant negative correlation with the EXG index, a significant negative correlation with the R, r, RBRI, and IKAW indices, and a significant positive correlation with GRRI, GRVI, and MGRVI. On the other hand, DMA, PN, GN, TKW, and Y did not show any significant correlation with any of the digital image indices (Figure 2a).

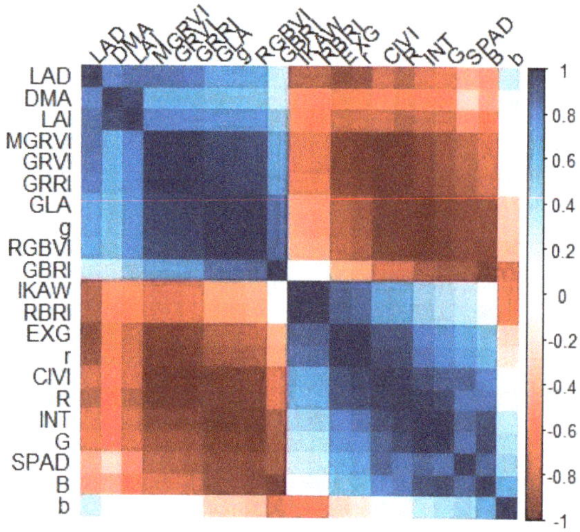

(a)

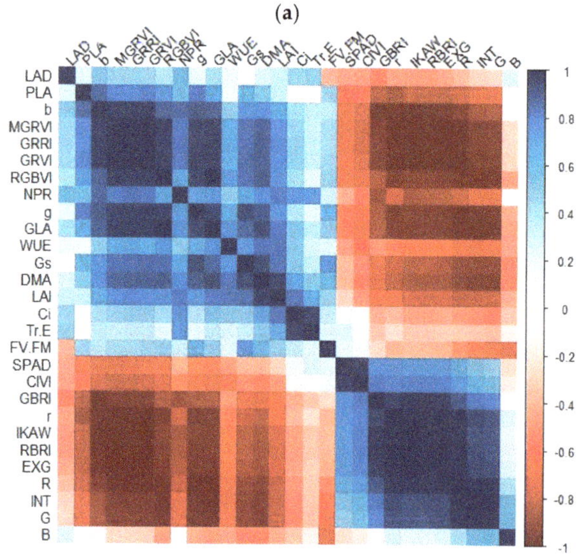

(b)

Figure 2. Correlation coefficients of different digital image indices with wheat physiological indicators at jointing and flowering stages. (**a**) Jointing stage. (**b**) Flowering stage.

A highly significant positive correlation was observed between the SPAD of wheat during the flowering stage (6) and the CIVE index (Figure 2b). The single plant leaf area (PLA) showed a significant positive correlation with the g and b indices and a significant negative correlation with the r, EXG, and GBRI indices. The correlation analysis revealed that the leaf area index (LAI) was positively correlated with the green leaf area index (GLA). On the other hand, the dry matter content (DMA) showed a highly significant positive correlation with the red (R), green (G), and near-infrared (INT) indices and a highly significant negative correlation with the GLA index. Furthermore, DMA exhibited significant negative correlations with the red edge (r), excess green (EXG), red–blue ratio index (RBRI), and inverse difference vegetation index (IKAW) and significant positive correlations with the green (g), green–red ratio index (GRRI), modified green–red index (MGRVI), and red–green–blue index (RGBVI). Stomatal conductance had a negative correlation with the G and INT indices, and a positive correlation with the g index. Additionally, stomatal conductance had negative correlations with the R, EXG, RBRI, and IKAW indices, and positive correlations with GRRI, GRVI, MGRVI, RGBVI, and GLA, all of which showed high or significant levels of correlation. In contrast, the net photosynthetic rate showed positive correlations with the b, GRRI, GRVI, and MGRVI indices and negative correlations with the EXG, GBRI, RBRI, and IKAW indices. However, the fluorescence parameters, intercellular CO_2 concentration, transpiration rate, and water use efficiency did not show significant correlations with any of the digital image indices.

3.2. *Relationships between Digital Image Indices and Wheat Physiological Indicators at the Time of Filling and Maturing*

During the wheat filling stage (7), the correlations between the wheat physiological indicators SPAD, LAI, LAD, and DMA and each digital image index were analyzed. SPAD had a highly significant negative correlation with the EXG and CIVE indices and a highly significant positive correlation with the GRRI, GRVI, and MGRVI indices. SPAD had significant positive correlations with the g, RGBVI, and GLA indices and significant negative correlations with the R, G, and INT indices (see Figure 3a). In addition, LAI had significant negative correlations with the R, G, INT, and CIVE indices and significant positive correlations with the g and GLA indices. Conversely, the LAD showed significant negative correlations with the RBRI and IKAW indices and a significant positive correlation with the b index. Similarly, the DMA had significant negative correlations with the EXG and CIVE indices and significant positive correlations with the GRRI, GRVI, and MGRVI indices.

At the wheat maturity stage (8), physiological indices, such as DMA and plant nitrogen content, were correlated with each digital image index in Figure 3b. The analysis revealed significant positive correlations between DMA and the g, GBRI, RBRI, IKAW, RGBVI, and GLA indices and significant negative correlations with the R, G, B, b, INT, and CIVE indices. No significant correlations were found between the plant nitrogen content (NC), stem and leaf nitrogen content (SLNC), stem and leaf phosphorus content (SLPC), kernel nitrogen content (KNC), kernel phosphorus content (KPC), glume nitrogen content (GNC), glume phosphorus content (GPC), number of spikes (PN), number of grains in a spike (GN), thousand kernel weight (TKW), and yield (Y) and any of the digital image indices. The stem and leaf potassium content (SLKC) had significant negative correlations with the R, G, B, and INT indices. Additionally, the seed potassium content (KKC) showed significant negative correlations with the g, GBRI, RGBVI, and GLA indices but significant positive correlations with the b and CIVE indices. Furthermore, the glumes' potassium content (GKC) showed significant positive correlations with the g, GBRI, GRVI, MGRVI, RGBVI, and GLA indices but significant negative correlations with the EXG and CIVE indices.

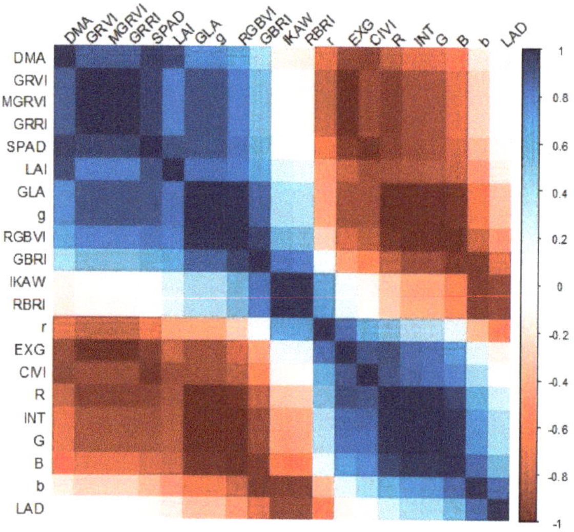

(a)

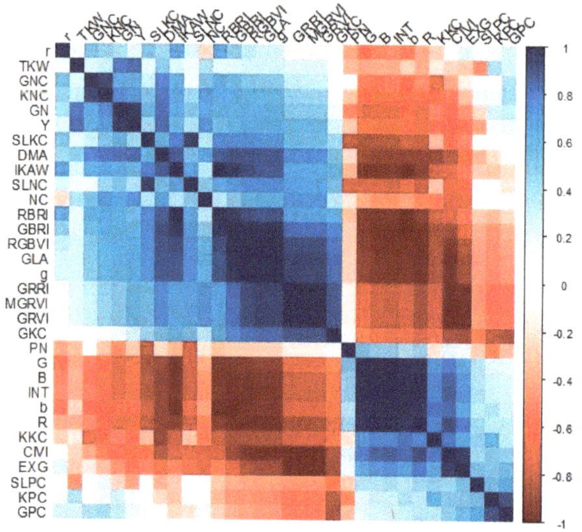

(b)

Figure 3. Correlation coefficients of different digital image indices with wheat physiological indicators at filling and mature stages. (**a**) Filling stage. (**b**) Mature stage.

3.3. Changes in Digital Image Indices during Various Fertile Stages of Wheat under Fertilizer Treatments

Figure 4 shows the differences and variations over time in wheat digital image indices across different fertilization treatments. The values of each digital image index were higher in the CK and CF of fertilization treatments compared to the other fertilization treatments. The index B of the digital image showed an increasing trend followed by a decreasing trend in the CF-P0 and CF-K0 fertilization treatments, while it displayed a decreasing trend followed by an increasing trend in other fertilization treatments. On the other hand, the digital image index INT showed a more gradual change in the CF-P0 and CF-K0 fertilization treatments, while it displayed a distinct trend of decreasing followed by increasing in other fertilization treatments. Additionally, the index of INT, which represents the digital image, changed at a slower rate in the CF-P0 and CF of fertilization treatments. In contrast, it exhibited a relatively stable trend in the CF-P0 and CF-K0 fertilization treatments. However, in other fertilization treatments, it displayed a trend of initially decreasing followed by an increase.

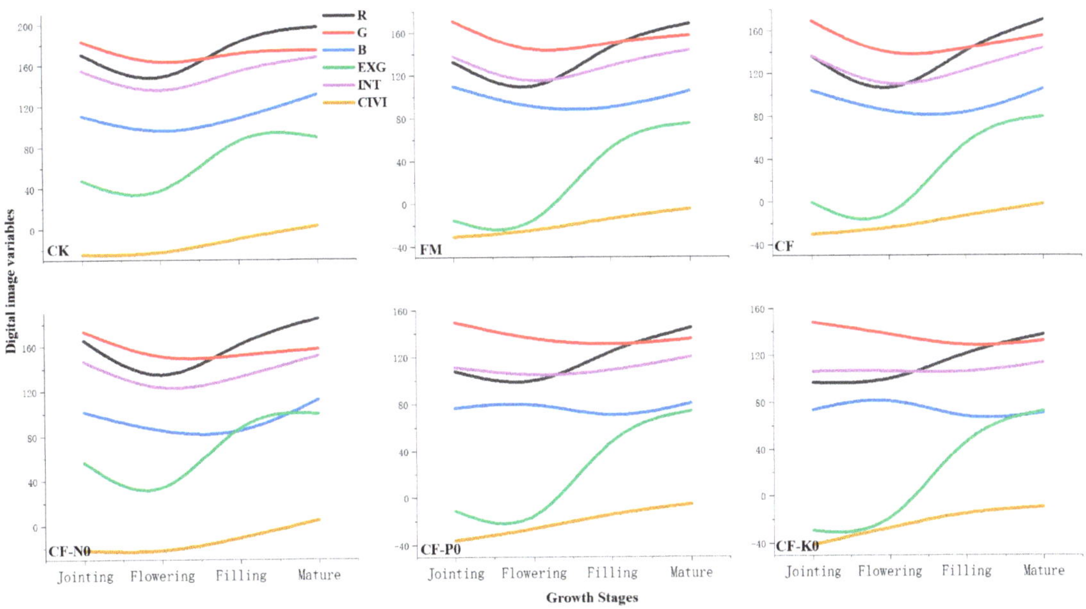

Figure 4. Variation in wheat digital image indices with fertility period under different fertilization treatments.

The digital image index GRRI showed a low value in the CK of the fertilization treatment and a relatively higher value in treatments with nitrogen application. The digital image indices GBRI and RBRI displayed an increasing trend with fertility in the CK of the fertilization treatment while showing a decreasing–increasing–decreasing trend in treatments with nitrogen application. The RBRI digital image index underwent a slight change in the CK of the fertilization treatment and a relatively significant change in treatments with nitrogen application (Figure 5). Figure 6 shows that the digital image indices GRVI and MGRVI had lower values in the CK of the fertilization treatment but relatively higher values in treatments with nitrogen application. Both indices exhibited a trend of increasing and then decreasing in correlation with the fertility process. The digital image index IKAW underwent a minor change in the CK of the fertilization treatment but a relatively significant change in treatments with nitrogen application. The other digital image indices did not show sensitivity to the nitrogen application treatments.

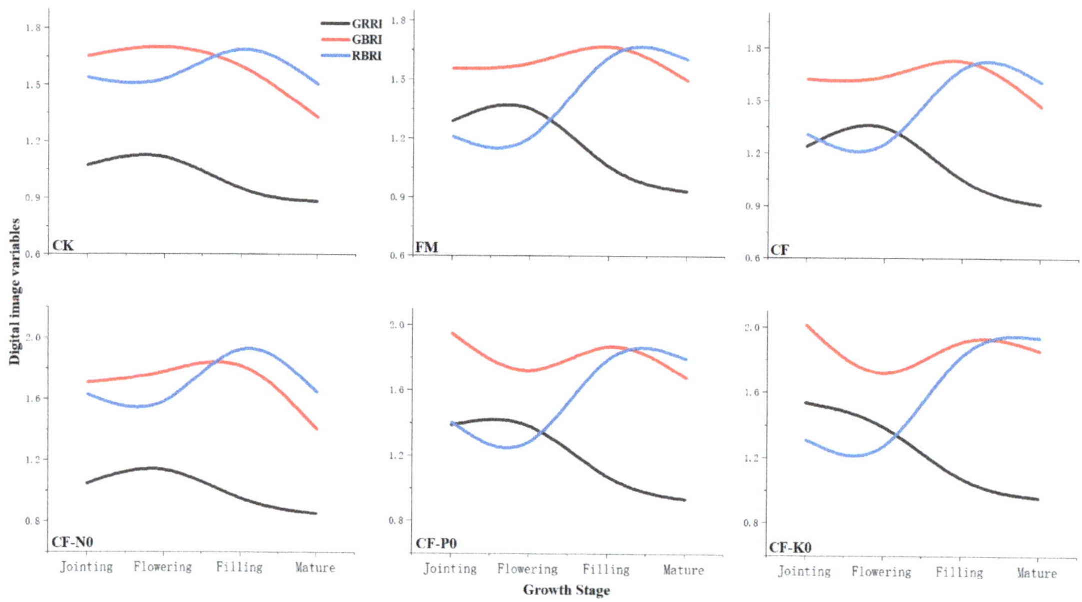

Figure 5. Variation in wheat digital image indices with fertility period under different fertilization treatments.

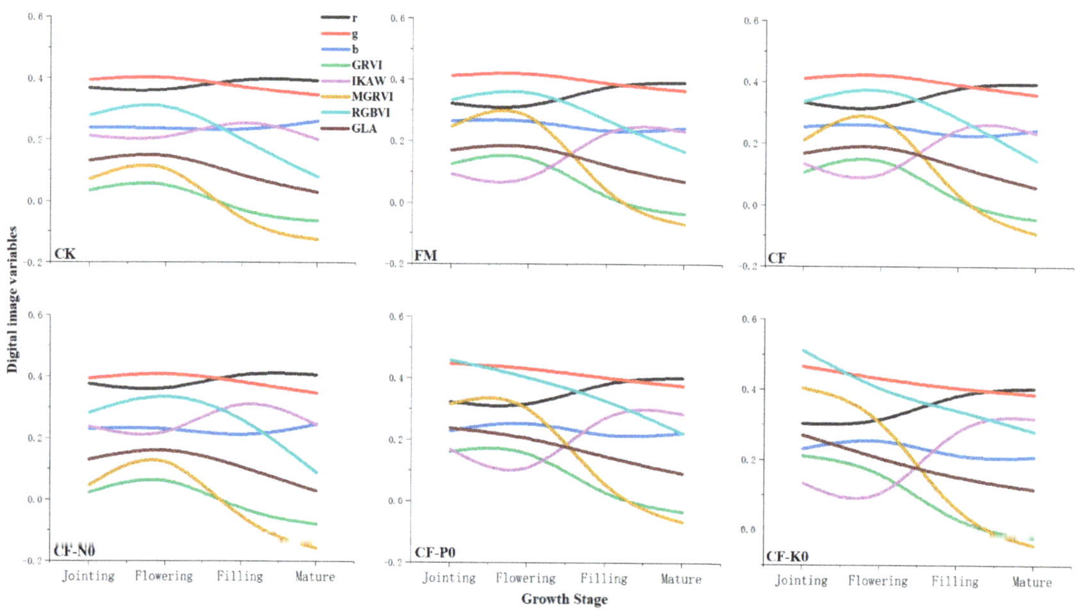

Figure 6. Variation in wheat digital image indices with fertility period under different fertilization treatments.

4. Discussion

Hyperspectral and multispectral remote sensing bands provide abundant and uninterrupted information, enabling precise feature recognition and mapping. However, their use in field production has been limited due to the high cost of equipment, their complex operation, and high technological requirements. Since the early 21st century, coinciding with the widespread use of digital cameras, research into crop growth monitoring using digital images has significantly increased. However, the use of handheld or stationary digital cameras has been limited by low image resolution and restricted indices in digital images, preventing this research from evolving into a self-contained system. On the other hand, UAV digital images have become increasingly popular for small-to-medium-scale crop monitoring in farmlands due to their straightforward operation, relatively low cost, and minimal technical requirements. The DJI MAVIC 2 Professional UAV was chosen for this study because of its widespread use in contemporary applications and its technical parameters and application effects that are ideally suited to meet the requirements of crop production monitoring in research areas.

This study used UAV remote sensing to capture digital images of spring wheat during four crucial fertility stages. The commonly used digital image indices of the crop were analyzed to establish quantitative relationships between these indices and the physiological indicators of wheat during each fertility period. The study aimed to investigate the potential of UAV digital images for monitoring the growth of spring wheat under various fertilization treatments. Based on the correlation analysis results presented in Figure 7, the digital image indices GLA, R, G, INT, g, GRVI, MGRVI, RGBVI, EXG, and GRRI were found to have a higher frequency of significant correlation with physiological indices during the four key reproductive periods of wheat. Therefore, they are considered more suitable for characterizing the physiological indices of wheat. The indices of r and CIVE have a low correlation with physiological indicators across wheat reproductive periods and are not recommended for use in monitoring wheat physiological indicators. The same results have been found in other studies as well [16,17]. EXG, MGRVI, and RGBVI consistently showed strong correlations with biomass and yield across all fertility stages of winter wheat [7]. The present study showed that these indices also exhibited significant correlations with biomass during the flowering period of spring wheat. However, their performance was not exceptional in other fertility periods.

Xiao [18] found a significant correlation between the aboveground biomass of winter wheat during the pulling stage and several digital image indices, including R, G, B, GRRI, GBRI, RBRI, r, g, and b. However, in the present study, no significant correlation was found between the aboveground biomass at the jointing stage of spring wheat and any of the digital image indices. Both studies found a significant correlation between the digital image indices R, G, and GBRI and changes in SPAD. The results of this study indicate that digital image indices R, G, and B exhibit significant negative correlations with biomass at all fertility periods in spring wheat. Similar findings were observed in studies on rice, maize, and spring wheat [18–20]. Umut [21] concluded that the digital image indices RGBVI and GLA were significantly correlated with the leaf area index (LAI) in spring wheat. Similarly, the present study found a significant positive correlation between the digital image index GLA and LAI in spring wheat at filling and mature stages. However, RGBVI did not exhibit a significant correlation with LAI. The study found a strong correlation between digital image indices and physiological indices of wheat during the flowering stage. The mean correlation coefficients were flowering (6) (0.638) > filling (7) (0.626) > maturity (8) (0.548) > jointing (3) (0.523). In contrast, Yang [7] found that the correlation coefficients for winter wheat digital image indices with biomass and yield at different fertility periods followed the order of jointing, flowering, tasseling, and pre-overwintering. The difference in the above results may be due to variations in wheat planting times. The ANOVA results indicate that the various fertilization treatments had significant differential effects only on the digital image indices G, B, and INT. The colors in the UAV images appeared more saturated during the wheat flowering stage compared to the jointing stage; this is mainly due to the fact that the indices lose sensitivity at this stage [22–24]. During the later stages of wheat fertility, the digital image index showed a significant correlation with potassium content in all parts of the plant. However, there was no significant correlation with nitrogen and phosphorus. These findings are consistent with a study conducted by Shi and Wang [25,26]. While several representative and complementary imaging indices and indicators were screened in this study, the current study has certain limitations in its selection of digital image indices and physiological indicators. Future research will seek to expand and enrich the relevant aspects examined in this study.

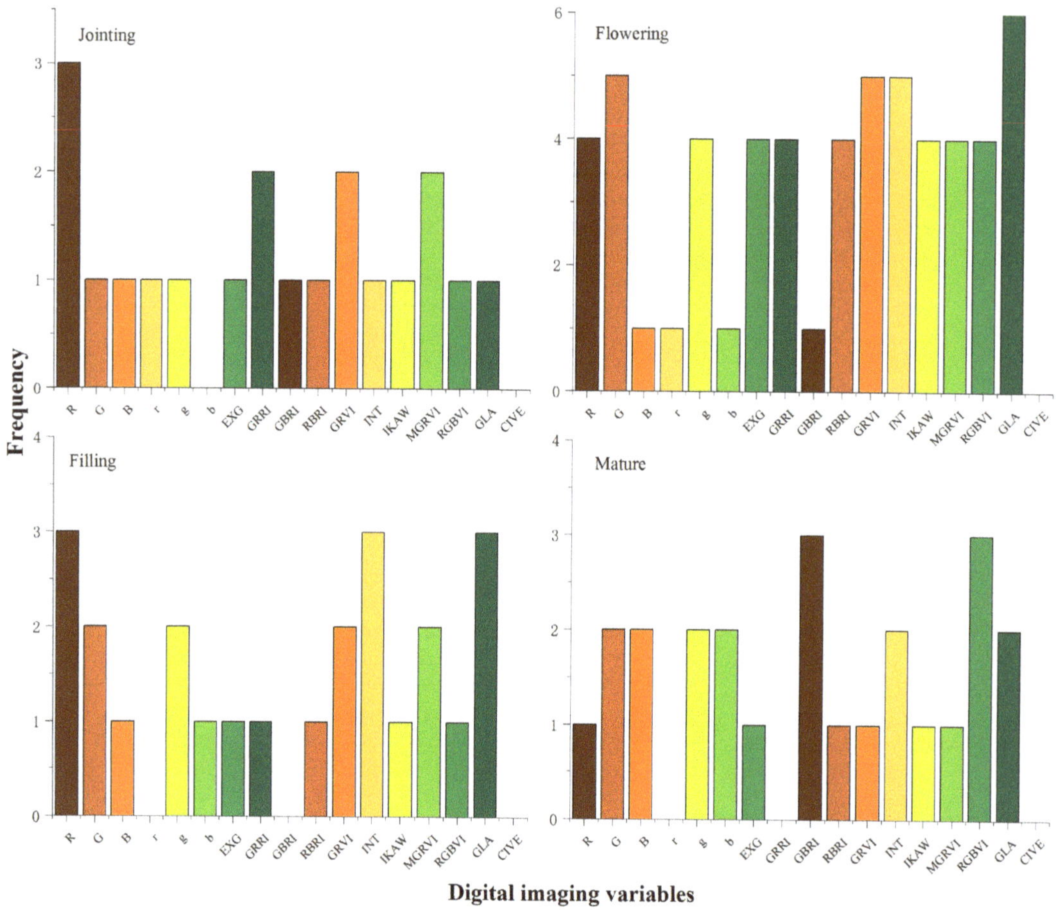

Figure 7. Frequency of significant correlation between each digital image index and physiological indicators at different wheat reproductive periods.

5. Conclusions

UAV digital images have become increasingly popular for small-to-medium-scale crop monitoring in farmlands due to their straightforward operation, relatively low cost, and minimal technical requirements. This advancement in unmanned aerial vehicle (UAV) digital image monitoring technology and methods will enable its application to other crops and further promote the development of UAV remote sensing monitoring. The DJI MAVIC 2 Professional UAV was chosen for this study because of its widespread use in contemporary applications and its technical parameters and application effects that were ideally suited to meet the requirements of crop production monitoring in the research areas. On the other hand, several digital image indices, such as GLA, R, G, INT, g, GRVI, MGRVI, RGBVI, EXG, and GRRI, showed significant correlations with physiological indices during the four primary reproductive stages of wheat. Specifically, these digital image indices could effectively characterize the physiological indices of spring wheat during the flowering stage (6). The results of this experiment showed that UAV digital images could effectively carry out nondestructive inspections of crops and provide a new way to monitor the growth status of wheat in the river-loop area.

Author Contributions: M.X.: writing—original draft. Y.Z.: supervision. J.L.: investigation. L.L.: investigation. P.Z.: formal analysis. Q.W.: visualization. M.L.: resources. H.W.: investigation. All authors have read and agreed to the published version of the manuscript.

Funding: This research was funded by the Inner Mongolia Natural Science Foundation Program, grant number "2021MS03089", the Inner Mongolia Basic Research Operating Expenses Colleges and Universities Program, grant number "BR231509", and the Inner Mongolia Autonomous Region Higher Education Institutions Carbon Peak Carbon Neutral Research Special Project, grant number "STZX202214".

Institutional Review Board Statement: Not applicable.

Informed Consent Statement: Not applicable.

Data Availability Statement: The data provided in this study are available upon request from the corresponding author. The data are not publicly available due to privacy.

Acknowledgments: This research was supported by the Inner Mongolia's innovative extension system for promoting and innovating the wheat industry.

Conflicts of Interest: The authors declare no conflicts of interest.

References

1. Omia, E.; Bae, H.; Park, E.; Kim, M.S.; Baek, I.; Kabenge, I.; Cho, B.-K. Remote Sensing in Field Crop Monitoring: A Comprehensive Review of Sensor Systems, Data Analyses and Recent Advances. *Remote Sens.* **2023**, *15*, 354. [CrossRef]
2. Li, W.; Wang, K.; Han, G.; Wang, H.; Tan, N.; Yan, Z. Integrated diagnosis and time-series sensitivity evaluation of nutrient deficiencies in medicinal plant (*Ligusticum chuanxiong* Hort.) based on UAV multispectral sensors. *Front. Plant Sci.* **2023**, *13*, 1092610. [CrossRef] [PubMed]
3. Benincasa, P.; Antognelli, S.; Brunetti, L.; Fabbri, C.A.; Natale, A.; Sartoretti, V.; Modeo, G.; Guiducci, M.; Tei, F.; Vizzari, M. Reliability of NDVI derived by high resolution satellite and UAV compared to in-field methods for the evaluation of early crop N status and grain yield in wheat. *Exp. Agric.* **2018**, *54*, 604–622. [CrossRef]
4. Corti, M.; Cavalli, D.; Cabassi, G.; Vigoni, A.; Degano, L.; Gallina, P. Application of a low-cost camera on a UAV to estimate maize nitrogen-related variables. *Precis. Agric.* **2019**, *20*, 675–696. [CrossRef]
5. Mullan, D.J.; Reynolds, M. Quantifying genetic effects of ground cover on soil water evaporation using digital imaging. *Funct. Plant Biol.* **2010**, *37*, 703–712. [CrossRef]
6. Sharma, R.; Kumar, M.; Alam, M.S. Image processing techniques to estimate weight and morphological parameters for selected wheat refractions. *Sci. Rep.* **2021**, *11*, 20953. [CrossRef] [PubMed]
7. Yang, J.; Ding, F.; Chen, C.; Liu, T.; Sun, C.; Ding, D.; Huo, Z. Correlation of wheat biomass and yield with UAV image characteristic parameters. *Trans. Chin. Soc. Agric. Eng.* **2019**, *35*, 104–110.
8. Poley, L.G.; Mcdermid, G.J. A Systematic Review of the Factors Influencing the Estimation of Vegetation Aboveground Biomass Using Unmanned Aerial Systems. *Remote Sens.* **2020**, *12*, 1052. [CrossRef]
9. Wang, X.; Wang, M.; Wang, S.; Wu, Y. Extraction of vegetation information from visible unmanned aerial vehicle images. *Trans. Chin. Soc. Agric. Eng.* **2015**, *31*, 152–159.
10. Costa, L.; Kunwar, S.; Ampatzidis, Y.; Albrecht, U. Determining leaf nutrient concentrations in citrus trees using UAV imagery and machine learning. *Precis. Agric.* **2022**, *23*, 854–875. [CrossRef]
11. Milas, A.S.; Romanko, M.; Reil, P.; Abeysinghe, T.; Marambe, A. The importance of leaf area index in mapping chlorophyll content of corn under different agricultural treatments using UAV images. *Int. J. Remote Sens.* **2018**, *39*, 5415–5431. [CrossRef]
12. Feng, D.; Xu, W.; He, Z.; Zhao, W.; Yang, M. Advances in plant nutrition diagnosis based on remote sensing and computer application. *Neural Comput. Appl.* **2020**, *32*, 16833–16842. [CrossRef]
13. Xia, S.S.; Zhang, C.; Li, J.Z.; Li, H.J.; Zhang, Y.M.; Hu, C.S. Diagnosis of nitrogen nutrient and recommended fertilization in summer corn using leaf digital images of cellphone camera. *Chin. J. Eco-Agric.* **2018**, *26*, 703–709.
14. Wang, Y.; Wang, D.J.; Zhang, G.; Wang, C. Digital camera based image segmentation of rice canopy and diagnosis of nitrogen nutrition. *Trans. Chin. Soc. Agric. Eng.* **2012**, *28*, 131–136.
15. Mao, W.; Wang, Y.; Wang, Y.R. *Real-Time Detection of Between-Row Weeds Using Machine Vision*; ASABE: St. Joseph, MI, USA, 2003.

16. Niu, Y.; Zhang, L.; Zhang, H.; Han, W.; Peng, X. Estimating Above-Ground Biomass of Maize Using Features Derived from UAV-Based RGB Imagery. *Remote Sens.* **2019**, *11*, 1261. [CrossRef]
17. Guo, T.; Yan, A.; Geng, H. Prediction of Wheat Plant Height and Leaf Area Index Based on UAV Image. *J. Triticeae Crops* **2020**, *40*, 1129–1140.
18. Xiao, Y.B.; Jia, L.L.; Chen, X.P.; Zhang, F. N status diagnosis of winter wheat by using digital image analysis technology. *Chin. Agric. Sci. Bull.* **2008**, *24*, 448–453.
19. Zhou, H.-J.; Liu, Y.-D.; Fu, J.; Sui, F.-G.; Cui, R. Analysis of Maize Growth and Nitrogen Nutrition Status Based on Digital Camera Image. *J. Qingdao Agric. Univ. (Nat. Sci.)* **2015**, *32*, 1–7.
20. Jia, L.-L.; Fan, M.-S.; Zhang, F.-S.; Chen, X.-P.; Lü, S.-H.; Sun, Y. Nitrogen Status Diagnosis of Rice by Using a Digital Camera. *Spectrosc. Spectr. Anal.* **2009**, *29*, 2176–2179.
21. Umut, H. Wheat Leaf Area Index (LAI) Inversion by Using "Satellite-UAV-Ground" Multi-Source Remote Sensing Data. Ph.D. Thesis, Xinjiang University, Urumqi, China, 2019.
22. Li, W.; Jiang, J.; Weiss, M.; Madec, S.; Tison, F.; Philippe, B.; Comar, A.; Baret, F. Impact of the reproductive organs on crop BRDF as observed from a UAV. *Remote Sens. Environ.* **2021**, *259*, 112433. [CrossRef]
23. Xu, L.; Zhou, L.; Meng, R.; Zhao, F.; Lv, Z.; Xu, B.; Zeng, L.; Yu, X.; Peng, S. An improved approach to estimate ratoon rice aboveground biomass by integrating UAV-based spectral, textural and structural features. *Precis. Agric.* **2022**, *23*, 1276–1301. [CrossRef]
24. Zhang, Z.; Flores, P.; Igathinathane, C.; Igathinathane, C.; Naik, D.; Kiran, R.; Ransom, J.K. Wheat lodging detection from UAS imagery using machine learning algorithms. *Remote Sens.* **2020**, *12*, 1838. [CrossRef]
25. Shi, X.-Y. Evaluation of Some Physiological Parameters of Winter Wheat Leaves Using Digital Image Processing. Ph.D. Thesis, Heibei Agriculture University, Baoding, China, 2005.
26. Wang, Y.; Wang, D.; Shi, P.; Omasa, K. Estimating rice chlorophyll content and leaf nitrogen concentration with a digital still color camera under natural light. *Plant Methods* **2014**, *10*, 36. [CrossRef]

applied sciences

MDPI

Article

The Effects of Vermicompost and Steel Slag Amendments on the Physicochemical Properties and Bacterial Community Structure of Acidic Soil Containing Copper Sulfide Mines

Xiaojuan Wang [1], Jinchun Xue [1,*], Min He [2,*], Hui Qi [1] and Shuting Wang [1]

[1] School of Energy and Mechanical Engineering, Jiangxi University of Science and Technology, Nanchang 330013, China
[2] School of Software Engineering, Jiangxi University of Science and Technology, Nanchang 330013, China
* Correspondence: xuejinchun@jxust.edu.cn (J.X.); 13653791569@163.com (M.H.)

Abstract: Acidification and heavy metal stress pose challenging threats to the terrestrial environment. This investigation endeavors to scrutinize the combined effects of vermicompost and steel slag, either singularly or in concert with Ryegrass (*Lolium perenne* L.), on the remediation of acidic soil resulting from sulfide copper mining. The findings illuminate substantial ameliorations in soil attributes. The application of these amendments precipitates an elevation in soil pH of 1.39–3.08, an augmentation in organic matter of 4.05–8.65, a concomitant reduction in total Cu content of 43.2–44.7%, and a marked mitigation in Cu bioavailability of 64.2–80.3%. The pronounced reduction in soil Cu bioavailability within the steel slag treatment group (L2) is noteworthy. Characterization analyses of vermicompost and steel slag further elucidate their propensity for sequestering Cu^{2+} ions in the soil matrix. Concerning botanical analysis, the vermicompost treatment group (L1) significantly enhances soil fertility, culminating in the accumulation of 208.35 mg kg^{-1} of Cu in *L. perenne* stems and 1412.05 mg kg^{-1} in the roots. Additionally, the introduction of vermicompost and steel slag enriches soil OTU (Operational Taxonomic Units) quantity, thereby augmenting soil bacterial community diversity. Particularly noteworthy is the substantial augmentation observed in OTU quantities for the vermicompost treatment group (L1) and the combined vermicompost with steel slag treatment group (L3), exhibiting increments of 126.04% and 119.53% in comparison to the control (CK). In summation, the application of vermicompost and steel slag efficaciously diminishes the bioavailability of Cu in the soil, augments Cu accumulation in *L. perenne*, induces shifts in the soil microbial community structure, and amplifies soil bacterial diversity. Crucially, the concomitant application of vermicompost and steel slag emerges as a holistic and promising strategy for the remediation of sulfide copper mining acidic soil.

Keywords: vermicompost; steel slag; copper; soil remediation; bacterial diversity

1. Introduction

Mining activities have precipitated extensive ecological degradation and environmental contamination, emerging as a paramount global concern. Due to the sulfur content inherent in ore constituents, sulfide copper ores undergo insufficient separation during ore beneficiation processes. Unseparated sulfur and tailings are collectively deposited in waste piles, culminating in long-term soil exposure and pervasive issues of soil acidification and heavy metal contamination [1]. Furthermore, excessive copper (Cu) levels in the soil incite plant toxicity symptoms, severely compromising plant establishment and growth [2]. Hence, the judicious selection of efficient and cost-effective amendments for the remediation of acidic soils associated with sulfide copper mining assumes pivotal significance. Establishing a secure and stable vegetation ecosystem is of paramount importance for the ecological restoration of soils in metal-contaminated mining areas.

In recent years, a plethora of inorganic and organic amendments has found widespread application in the remediation of soils tainted with heavy metals, constituting what is known as in situ immobilization remediation. By judiciously adding amendments to the soil, this technique modulates the form of heavy metals, reducing their mobility and bioavailability. This presents an effective, cost-efficient, and sustainable remediation paradigm [3]. However, inherent disparities in the composition of various amendments may impart corresponding limitations to their efficacy in soil remediation. Notably, biochar, endowed with a meticulously defined porous structure and an abundance of oxygen-containing functional groups, exhibits commendable adsorption capabilities for soil pollutants. Nevertheless, uncertainties enveloping the aging of biochar may portend potential environmental risks, and its exorbitant cost may render it impracticable for large-scale mining restoration initiatives [4]. While the application of lime expeditiously elevates the pH of acidic soils, its aptitude to sustain soil pH equilibrium is circumscribed, rendering it susceptible to secondary acidification and potential soil compaction with protracted use [5]. Ergo, the quest for environmentally friendly and cost-effective amendments stands as a linchpin in the realization of in situ immobilization remediation of soils. Extant research posits that vermicompost, replete with organic matter and diverse amino acids, proffers superlative water retention, breathability, high fertility, and enduring fertility. They efficaciously ameliorate the physical, chemical, and biological properties of the soil. Additionally, vermicompost, with its discernibly porous structure, can serve as adept adsorbents and passivators for heavy metals [6]. As a byproduct of the steel production process, steel slag has a utilization rate of only approximately 22%. Steel slag not only burnishes soil structure and regulates soil pH but also enhances the soil's proclivity for metal ions, fomenting the formation of hydroxides. Ergo, employing steel slag as a soil amendment for the remediation of acidic soils associated with sulfide copper mining is a sagacious and sustainable choice [7]. *L. perenne,* with its potential for phytostabilization of acidic and heavy metal-contaminated soils, emerges as an eminently suitable candidate [8]. The concomitant cultivation of *L. perenne* affords a holistic evaluation of the remediation effects of vermicompost and steel slag on acidic soils associated with sulfide copper mining.

The ultimate aim of soil remediation transcends the mere reduction in the bioavailability and mobility of deleterious metals within the soil; it is paramount to elevate the health indices of the soil. The intricacies of the soil microbial community structure wield profound importance in augmenting soil functionality, representing a pivotal metric for assessing soil vitality. Consequently, a heightened focus on the nuanced alterations within soil microbial diversity becomes imperative [9]. A compendium of scholarly investigations attests that the introduction of ameliorative agents serves to augment the diversity inherent in the soil microbial framework, thereby amplifying the functional repertoire of these microorganisms. Notably, the application of steel slag manifests a conspicuous augmentation within a specific cohort of soil bacterial communities, bestowing considerable utility upon soil–plant remediation endeavors [10]. Correspondingly, vermicompost furnishes a plethora of nutrients conducive to microbial proliferation, thereby potentiating heightened microbial activity and functionality. This augmentation, in turn, engenders an enhancement in microbial respiration, culminating in a more efficacious amelioration of soil structure [11]. It is noteworthy that the extant body of literature is somewhat bereft in delving into the collective impact of vermicompost and steel slag on the structural dynamics of soil microbial communities in the co-presence of vegetation.

This investigation employs vermicompost and steel slag as soil amendments, utilizing them individually or in conjunction with *L. perenne* for the amelioration of acidic soils associated with sulfide copper mining. The outlined objectives of this research endeavor are (1) to elucidate the mechanistic intricacies of vermicompost and steel slag via comprehensive characterization analysis; (2) to scrutinize the ramifications of the combined application of vermicompost, steel slag, and *L. perenne* on the physicochemical attributes of the soil, concomitant with an examination of the bioavailability of copper (Cu); (3) to analytically ascertain the biomass and heavy metal accumulation of *L. perenne* under the influence of vermicompost and steel slag; and (4) to assess the intricate interplay between soil environmental factors and the nuanced diversity characterizing the soil bacterial community structure.

2. Materials and Methods

2.1. Tested Soil and Amendments

The soil was procured from Chengmenshan Copper Mine (115°48′32″ E, 29°41′26″ N), situated in the Chaisang District of Jiujiang City, Jiangxi Province, China. Employing a meticulous five-point sampling method, a homogeneous soil sample was extracted from the surface layer (5–30 cm). Employing the quartile method, the soil samples underwent a partition into two segments. One segment was transported to the laboratory, where it underwent air-drying, grinding, and sieving through a 2 mm nylon sieve for subsequent utilization in pot experiments. The second segment found its place of preservation in a −80 °C freezer to facilitate DNA extraction. The steel slag utilized in this experimental inquiry emanated from a local iron and steel establishment, while vermicompost was procured from Huizhou Nianhe Agricultural Technology Co., Ltd., Huizhou, China. The examination of the surface microphotographs of vermicompost and steel slag was conducted via Scanning Electron Microscopy (SEM, Zeiss EVO18, Tokyo, Japan). X-ray Diffraction (XRD, Malvern Panalytical Empyrean, San Jose, CA, USA) was instrumental in discerning the mineral composition of the steel slag. Fourier Transform Infrared Spectroscopy (FTIR, Thermo Scientific Nicolet iS20, Waltham, MA, USA) served as the analytical tool for probing the surface functional groups of vermicompost. The fundamental physicochemical parameters of the soil and amendments are presented in the following table (Table 1).

Table 1. Main physicochemical properties of the tested soil, vermicompost, and steel slag.

Properties	Tested Soil	V	S
pH	3.78	6.4	12.52
cation exchange capacity (CEC, cmol·kg^{-1})	17	na	na
soil organic matter (SOM, g·kg^{-1})	5.93	470.6	na
Total N (TN, g·kg^{-1})	0.98	16.94	na
Total P (TP, g·kg^{-1})	0.45	16.73	na
Total K (TK, g·kg^{-1})	5.61	12.88	na
Total Cu (g·kg^{-1})	2.41	na	0.02

Note: V, vermicompost; S, steel slag; na, valid values are not available.

2.2. Experimental Design

In March 2023, a potted experiment was conducted within the greenhouses of Jiangxi University of Science and Technology. Building upon the initial screening outcomes of our research group, a combination of vermicompost (4% *w*/*w*) and steel slag (2% *w*/*w*) was meticulously chosen as the amendment ratio for this investigation. White plastic pots featuring perforations at the base were filled with 2.5 kg of soil intermixed with the specified amendments. Preceding the commencement of the experiment, a thorough amalgamation of amendments and soil took place, followed by watering to achieve 70% of the maximum field water-holding capacity. A two-week equilibration period was observed. The experimental treatments comprised (1) the control soil devoid of any amendments (CK); (2) vermicompost (4% *w*/*w*) (L1); (3) steel slag (2% *w*/*w*) (L2); and (4) a combination of vermicompost (4% *w*/*w*) and steel slag (2% *w*/*w*) (L3). Each treatment was replicated three times.

L. perenne, selected for its remarkable capacity to accumulate heavy metals and manifest tolerance mechanisms to copper toxicity, served as the focal plant species in this study. Seeds of *L. perenne* were procured from the Yunguang flagship store. Fifty seeds were evenly distributed in each pot, and following germination, the seedlings were selectively thinned to maintain 30 plants per pot (all *L. perenne* within the CK treatment wilted and perished within two weeks under the duress of heavy metal stress). Daily irrigation was administered using deionized water, with soil moisture meticulously maintained at 70% of the maximum field water-holding capacity.

2.3. Plant Sampling and Analysis

After a six-month growth period, the *L. perenne* was harvested. The *L. perenne* was thoroughly washed with deionized water to remove any contaminants. The stems, leaves, and roots of the *L. perenne* were then separated and dried in an oven at 70 °C until a constant weight was achieved. The dry weight was recorded using a milligram balance. A 0.2 g dried plant sample was weighed in a conical flask and added with 5 mL of HNO_3 and 1 mL of $HClO_4$. The flask was placed on a graphite digestion device at 200 °C until the solution in the flask became colorless and transparent (this process took about 2–3 h, during which hydrogen peroxide can be added dropwise). The solution was then transferred into a 50 mL volumetric flask and diluted to the mark with distilled water. This is the extraction solution. The concentration of Cu in the extraction solution was directly determined via Inductively Coupled Plasma Optical Emission Spectrometry (ICP-OES) [12].

2.4. Soil Chemical Analysis

The soil samples were from the pots, and the plant residues were eliminated. The samples were passed through a 20 mm nylon sieve, and the corresponding chemical analyses were conducted [12]. The pH of the soil suspension was measured using a glass electrode with a water-to-soil ratio of 2.5:1. The soil electrical conductivity (EC) was determined using the conductivity method with a water-to-soil ratio of 5:1. The soil organic matter (SOM) was measured using the potassium dichromate dilution heat method [13]. The cation exchange capacity (CEC) was measured using the ammonium acetate method [14]. The soil redox potential (Eh) was measured using a platinum redox electrode (In Lab Redox). The total Cu content in the soil was determined via ICP-OES, and the available Cu content in the soil was determined via Diethylene Triamine Pentaacetic Acid (DTPA) extraction inductively coupled plasma emission spectroscopy [15].

2.5. Soil Bacterial Community Diversity Analysis

Genomic DNA from the microbial community was meticulously extracted utilizing the E.Z.N.A.® soil DNA Kit (Omega Bio-tek, Norcross, GA, USA), adhering to the manufacturer's prescribed instructions. The resultant DNA extract underwent scrutiny on a 1% agarose gel, and its concentration and purity were judiciously assessed using a NanoDrop 2000 UV-vis spectrophotometer (Thermo Scientific, Wilmington, NC, USA). The amplification of the bacterial 16S rRNA gene's hypervariable region V3-V4 was executed with the primer pairs 338F (5′-ACTCCTACGGGAGGCAGCAG-3′) and 806R (5′-GGACTACHVGGGTWTCTAAT-3′) employing an ABI GeneAmp® 9700 PCR thermocycler (ABI, Los Angeles, CA, USA). Subsequently, the purified amplicons were amalgamated in equimolar proportions and subjected to paired-end sequencing on an Illumina MiSeq PE300 platform or NovaSeq PE250 platform (Illumina, San Diego, CA, USA) in strict adherence to standardized protocols, as implemented by Majorbio Bio-Pharm Technology Co., Ltd. (Shanghai, China) [16].

2.6. Statistical Analysis

All statistical analyses were conducted using SPSS 26.0 software. Single-factor analysis of variance (ANOVA) and Duncan's multiple range test were employed for multiple comparisons (n = 3, $p < 0.05$). Graphs were created using Origin 2022 for data visualization. Additionally, symbiotic network graphs were generated using Gephi 0.9.2, and correlation heatmaps were constructed using R Studio 4.0.4.

3. Results and Discussion

3.1. Characterization of Amendments

3.1.1. SEM and FTIR Analysis of Vermicompost

SEM images vividly depict the intricately dispersed reticular architecture adorning the surface of vermicompost, presenting an exalted degree of porosity that affords a plethora of adsorption loci for the sequestration of heavy metals (Figure 1A). FTIR is an indispensable methodology for scrutinizing and evaluating of the molecular functional groups inherent in substances [17]. Via a discerning examination of the FTIR spectra of vermicompost pre- and postCu^{2+} adsorption (Figure 1B), a discernible congruity in the peak positions of the infrared spectra manifests itself, with the principal absorption bands aligning within the spectral range of 3300 to 3600 cm^{-1}. Notwithstanding, nuanced differentials in peak intensity are observed. The resonances at 3431 and 3425 cm^{-1} find attribution to the stretching vibrations of O-H in phenolic and alcoholic entities [18]. The zenith at 2928 cm^{-1} is ascribed to the stretching vibration of C-H within aliphatic moieties. The features at 1647 and 1651 cm^{-1} correlate with the stretching vibrations of C=C within aromatic compounds, serving as proxies for the aromatic compound content within vermicompost [19]. The apices at 1386 and 1412 cm^{-1} correspond with the deformation vibrations of C-H within -CH_3 and -CH_2 of aliphatic compounds [20]. Furthermore, the asymmetrical stretching vibrations of C-O within polysaccharides correlate with the peaks at 1041 and 1038 cm^{-1} [18]. In comparison to untreated vermicompost, a discernible attenuation in the intensity of the absorption band corresponding to the aromatic C=C bond is observed in vermicompost post-Cu^{2+} adsorption, imputed to the interaction between the delocalized π electrons inherent in aromatic structures and Cu^{2+} [21]. The absorption bands associated with O-H and C-O bonds also exhibit a diminution in intensity post-Cu^{2+} adsorption, with the primary impetus for this transformation potentially residing in the occurrence of chelation between Cu^{2+} and oxygen-bearing functional groups [22].

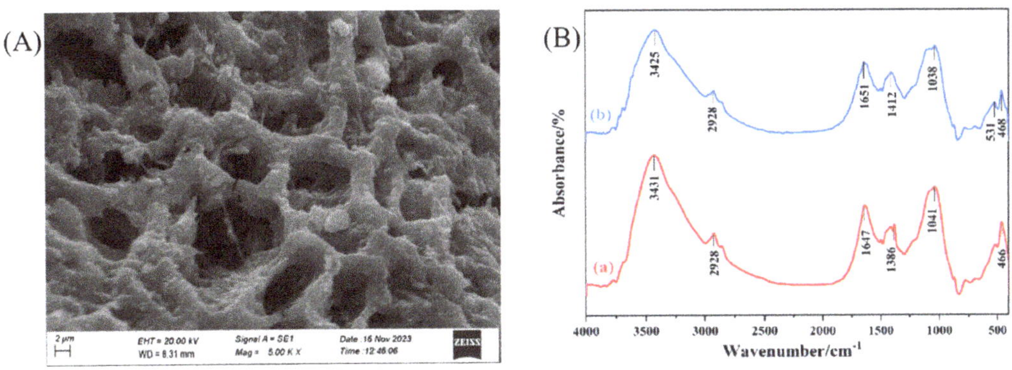

Figure 1. SEM images of vermicompost (**A**); FTIR spectra of vermicompost (a) and vermicompost after Cu^{2+} adsorption (b) (**B**).

3.1.2. SEM and XRD Analysis of Steel Slag

The SEM analysis illustrates (Figure 2A) the presence of numerous agglomerates on the surface of the steel slag, interspersed with a modicum of non-agglomerated entities, exhibiting an irregular disposition. A plethora of investigations affirms the stellar mesoporous adsorbent qualities of steel slag, characterized by robust affinities and efficacious removal rates for metallic ions [23]. The compositional scrutiny of steel slag, validated via XRD (Figure 2B), delineates predominant constituents in the form of oxygenated compounds of Fe, Ca, and Si, with preeminent phases encompassing FeO, $CaCO_3$, Ca_3SiO_5, Ca_2SiO_4, $Ca_2Fe_2O_5$, and SiO_2 [24]. The elevation in soil pH ostensibly ensues from the neutralizing influence of alkali compounds intrinsic to steel slag. Oxidative interactions of slag-borne substances with water yield -OH moieties, which subsequently engage in reactions with heavy metal cations within the soil milieu, precipitating hydroxides. In acidic soils replete with Cu^{2+}, a profusion of copper sulfide emerges via chemical precipitation reactions, culminating in the formation of $Cu(OH)_2$ [25,26]. Furthermore, Ca^{2+} within the steel slag engages in ion exchange with heavy metal ions in the soil, effectually securing heavy metals in situ [24]. It concurrently forms conglomerates with mineral colloids and organic constituents, thereby fostering a structural amelioration in acidic and heavy metal-afflicted soils [27].

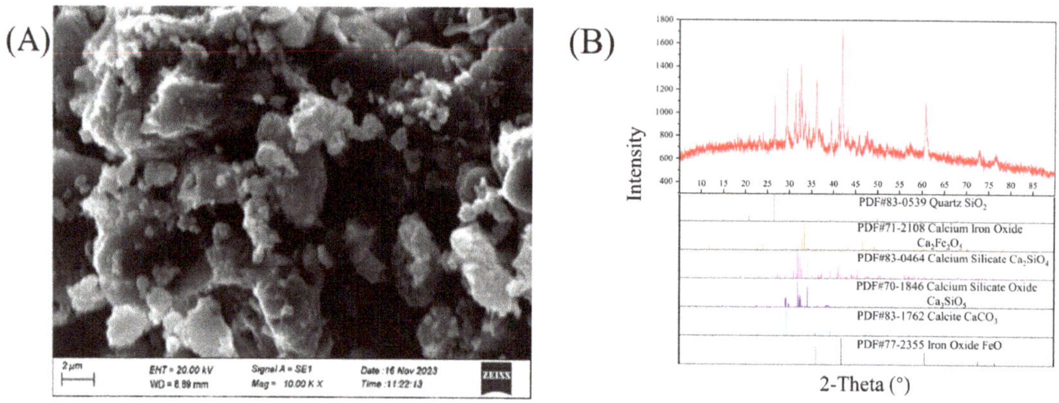

Figure 2. SEM images (**A**) and XRD patterns (**B**) of steel slag.

3.2. The Impact of Amendments on Soil Physicochemical Properties and the Bioavailability of Cu

The utilization of soil amendments has manifested improvements in soil physicochemical attributes to varying extents (Table 2). In contrast to the control (CK), the application of vermicompost and steel slag has exerted a salutary influence on soil pH. Notably, in the plot treated with vermicompost (L1), the pH has ascended from 3.78 to 5.17. In the plots ameliorated with steel slag (L2) and a composite of vermicompost and steel slag (L3), a more conspicuous augmentation in soil pH has been observed, reaching 6.76 and 6.86, respectively, from the baseline of 3.78. The escalation in soil pH is ascribed to the alkaline nature of steel slag, fostering the generation of hydroxides and the precipitation of carbonates [28]. Furthermore, the electrical conductivity (EC) within the soil across all treatment cohorts (L1, L2, and L3) has undergone diverse increments vis à vis CK, registering increments of 17.1%, 44.6%, and 34%, respectively. Concurrently, the cation exchange capacity (CEC) has undergone a notable surge, escalating from 17.2 cmol kg^{-1} to 30.6, 28.2, and 23.2 cmol kg^{-1}. Vermicompost and steel slag conduce to the augmentation of soil EC via the precipitation of soluble ions [29]. The heightened pH implies an accentuation of negative surface charges (OH^-), thereby instigating a heightened CEC [30]. Moreover, both the singular and concomitant application of amendments has signally elevated soil organic matter (SOM), particularly in the vermicompost-treated cohort (L1), where SOM has burgeoned by 147.9% compared to CK. The substantial augmentation in soil SOM is ostensibly attributable to the constituents bequeathed by vermicompost, with organic amendments serving as a direct fount for the amplification of SOM in the soil [31]. Intriguingly, post-amendment application, there was a marginal decrement in soil Eh. The inverse correlation between soil Eh and pH values substantiates this phenomenon [32].

Table 2. Basic physicochemical properties of soils under different treatments.

Treatments	pH	EC (mS cm^{-1})	SOM (g kg^{-1})	CEC (cmol kg^{-1})	Eh (mv)
CK	3.78 ± 0.45 c	858 ± 36.1 c	5.85 ± 0.05 d	17.2 ± 0.45 d	509 ± 6.06 a
L1	5.17 ± 0.17 b	1005 ± 73.4 bc	14.5 ± 0.27 a	30.6 ± 0.82 a	492 ± 3.38 b
L2	6.76 ± 0.04 a	1241 ± 47.2 a	9.90 ± 0.01 c	28.2 ± 0.26 b	417 ± 1.70 c
L3	6.86 ± 0.05 a	1150 ± 34.9 ab	12.5 ± 0.45 b	23.2 ± 0.51 c	403 ± 2.80 d

Note: different letters indicate obvious differences between treatment groups for the same index ($p < 0.05$). The data are mean ± SD, n = 3. EC, electrical conductivity; SOM, soil organic matter; CEC, cation exchange capacity; Eh, redox potential. CK: the control soil devoid of any amendment. L1: vermicompost amendment; L2: steel slag amendment; L3: a combination of vermicompost and steel slag amendment.

Due to the pronounced acidity inherent in the acidic soil derived from copper sulfide, characterized by a pH value of 3.78, the copper (Cu) within the soil manifests a heightened degree of bioavailability [33]. The application of vermicompost in isolation (L1), exclusive application of steel slag (L2), and the co-application of vermicompost and steel slag (L3) collectively yield a discernible reduction in both total and bioavailable Cu content within the soil matrix (Figure 3). Relative to the control group (CK), there is a notable decline in total Cu content of 43.2–44.7%, alongside an analogous decrease in bioavailable Cu content by 79.6–88.9%. Furthermore, the Cu bioavailability (Cu-BI) in soils subjected to L1, L2, and L3 applications exhibits a reduction of 64.2%, 80.3%, and 72.6%, respectively. These outcomes underscore the efficacy of steel slag application in enhancing Cu sequestration within the soil.

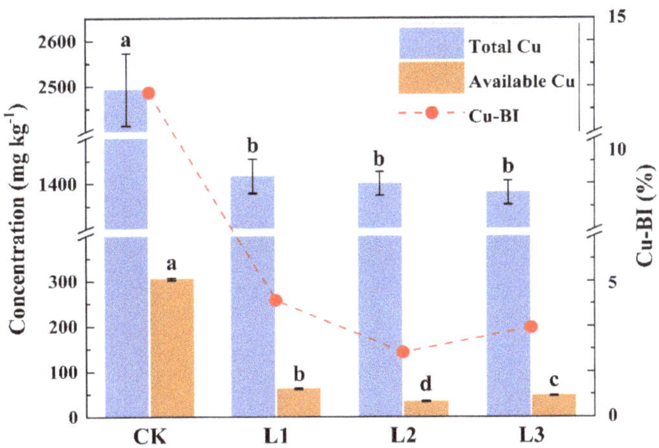

Figure 3. The total Cu concentration, available Cu concentration, and bioavailability of Cu in soil under different treatments. Note: The bars sharing the different letters suggest significant differences ($p < 0.05$). The data are mean ± SD, n = 3. CK: the control soil devoid of any amendment. L1: vermicompost amendment; L2: steel slag amendment; L3: a combination of vermicompost and steel slag amendment. The same as below.

The integration of SEM and XRD spectra unveils the substantial presence of alkaline oxides within the steel slag. This substantiates its robust acid-neutralizing capacity, thereby markedly elevating the concentration of hydroxide ions (OH^-) in the soil milieu. Consequently, this heightened OH^- concentration facilitates the precipitation of Cu^{2+} as insoluble $Cu(OH)_2$ [34]. Furthermore, SEM and FTIR analyses of vermicompost corroborate the abundance of organic functional groups. The undissociated functional groups form coordination bonds with liberated Cu^{2+} to promote the formation of less reactive metal complexes, thus significantly mitigating the toxicological impact of heavy metals [35].

3.3. Effects of Amendments on Biomass and Cu Bioaccumulation in *L. perenne*

In the preliminary experimental phase, it was discerned that the cultivation of black *L. perenne* in unmodified soil, particularly under the stringent conditions of acidic soil derived from copper sulfide, led to the curtailment of seed germination and the impediment of vegetative growth. The germination efficacy of black *L. perenne* was remarkably scant, culminating in comprehensive morbidity due to toxicity within a fortnight. Application of both vermicompost and steel slag, either independently or conjointly, evinced mitigating effects on the developmental trajectory of black *L. perenne*. The biomass of black *L. perenne* within the solitary vermicompost treatment cohort (L1) exhibited the nadir, while the amalgamated application of vermicompost and steel slag treatment group (L3) displayed the apogee in biomass (Figure 4A). Vermicompost and steel slag has been irrefutably substantiated to efficaciously enhance the physicochemical attributes of soil, endowing infertile soil with imperative nutrients, thereby catalyzing botanical proliferation and accentuating vegetative biomass [36–38]. The cumulative copper (Cu) accrual within the stems, leaves, and roots of black *L. perenne* adhered to the hierarchical sequence: the vermicompost treatment group (L1) surpassed the combined vermicompost and steel slag treatment group (L3), which in turn exceeded the steel slag treatment group (L2). In the vermicompost treatment group (L1), the cumulative Cu accumulation in the stems and roots of black *L. perenne* attained levels of 208.35 and 1412.05 mg kg^{-1}, respectively (Figure 4B). The attenuation in Cu accumulation observed in black *L. perenne* within the steel slag treatment group (L2) is attributable to the incorporation of steel slag, which precipitates an elevation in soil pH, consequently diminishing the bioavailability of Cu [39]. Within the vermicompost treatment group (L1), the substantial Cu accumulation in both the stems and roots of black *L. perenne* emanates from the diminished biomass, concomitant with a minor dilution effect. Additionally, the augmentation of soluble carbon in the soil via the application of vermicompost enhances Cu mobilization, facilitating its uptake by black *L. perenne* [26].

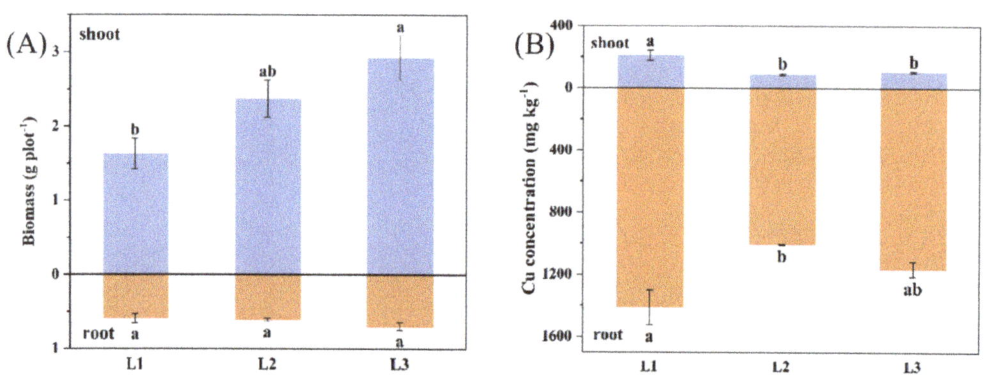

Figure 4. The biomass (**A**) and concentration of Cu (**B**) in *L. perenne* under different treatments. Note: the bars sharing the different letters suggest significant differences ($p < 0.05$).

3.4. Impact of Amendments on the Composition of Soil Bacterial Communities

Utilizing high-throughput sequencing methodologies, we analyzed the microbial community composition within soils subjected to diverse treatments. We derived results by employing a clustering threshold of 97% and normalizing to the minimum sample sequence count. The Venn diagram visually encapsulates the overlapping and distinct Operational Taxonomic Units (OTU) at the 97% similarity level across varied treatment modalities (Figure 5A). Common OTU shared among the control group (CK), the vermicompost treatment ensemble (L1), the steel slag treatment cohort (L2), and the confluence of vermicompost and steel slag treatment group (L3) were quantified at 65, with unique

OTU numbering 65, 42, 13, and 41, respectively. Relative to the control group, the addition of vermicompost and steel slag engendered a diminution in unique OTU within each treatment milieu, signifying an amelioration in the distinct bacterial taxa prevalent in acidic soil stemming from copper sulfide. Conversely, in comparison to the control group, all alternative treatment assemblages exhibited an augmentation in the aggregate number of bacterial OTU. Particularly noteworthy enhancements were discerned in the vermicompost treatment group (L1) and the combined vermicompost and steel slag treatment group (L3), showcasing an elevation of 126.04% and 119.53%, respectively.

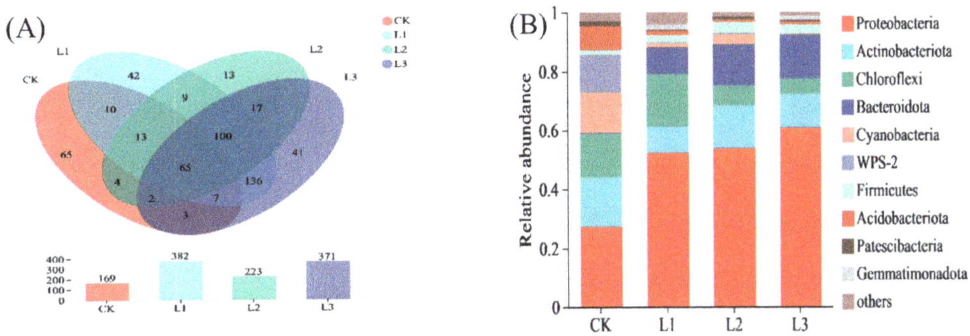

Figure 5. Venn diagram of soil bacteria at the OTU level (**A**) and the relative abundance of dominant bacterial communities at the phylum level (**B**) in soil under different treatments.

The taxonomic composition at the phylum level within distinct treatment cohorts exhibits congruity in predominant species but divergence in their relative abundances (Figure 5B). Broadly construed, the prevailing phyla in the bacterial consortium are *Proteobacteria* (27.48–60.94%), followed by *Actinobacteriota* (8.78–16.54%) and *Chloroflexi* (5.03–17.75%), collectively constituting over 60% of the bacterial assemblage. Antecedent investigations posit that Proteobacteria and *Chloroflexi* gradually ascend to the status of dominant species in soils contaminated with heavy metals, demonstrating formidable resilience to exigent environments [40,41]. In relation to the control group (CK), the augmentation of vermicompost and steel slag engenders an amplification in the prevalence of Proteobacteria, *Bacteroidota*, and Firmicutes while concurrently instigating a decrement in *Actinobacteriota*, WPS-2, and *Acidobacteriota*. Proteobacteria, characterized by heightened tolerance to pollutants, proficiently harness elevated concentrations of heavy metals and other deleterious substances as founts of energy and sustenance, thereby unveiling promising prospects for plant remediation and environmental safeguarding [42]. Furthermore, *Acidobacteriota* and *WPS-2*, ubiquitously distributed in desolate and exposed soil domains, undergo a marked attenuation in abundance within treatment groups subjected to amendments. This phenomenon is ostensibly attributed to the escalation in soil pH induced by the application of steel slag and vermicompost, effectually immobilizing Cu^{2+} [43–45].

3.5. The Relationship between Soil Physicochemical Properties and Bacterial Community

Envisaging the intricate interplay between soil physicochemical attributes and the intricate structure of bacterial communities, represented by the prevalence of ten prominent phyla, is achieved via the visualization derived from network analysis (Figure 6). Discerning from this analysis, a notably robust positive correlation emerges between Proteobacteria, *Bacteroidota*, and soil pH while concurrently revealing a pronounced inverse relationship with soil Eh and the aggregate Cu concentration. The soil's pH, recognized as a cardinal determinant, orchestrates a profound influence on the opulence of microbial communities therein [46]. The preeminent phylum, Proteobacteria, validated as an integral facilitator of plant flourishing, adept in the decomposition of soil macro-molecules and the facilitation of elemental cycling, surfaces as a key protagonist in this correlation [47]. *Acidobacteriota* manifests an unmistakable affirmative correlation with soil effective Cu concentration and an equally conspicuous negative association with soil EC. In stark contrast, Firmicutes, portraying a counteractive trend, exhibits a highly significant negative interrelation with soil effective Cu concentration and an inversely notable positive correlation with soil EC. This infers that the supplementation of vermicompost and steel slag serves to heighten the concentration of inorganic oxygen constituents such as Ca and Fe within the soil matrix, thereby exerting discernable effects on soil EC and effectively curtailing the bioavailability of Cu [48]. Concomitantly, soil SOM establishes a robustly affirmative correlation with *Gemmatimonadota*, juxtaposed against highly pronounced negative associations with *Actinobacteriota* and *Patescibacteria*. Scholarly investigations underscore the pivotal role of *Gemmatimonadota* in fostering vegetation recovery, with its abundance eliciting increasingly potent ameliorative effects on soil in tandem with rising soil fertility [49].

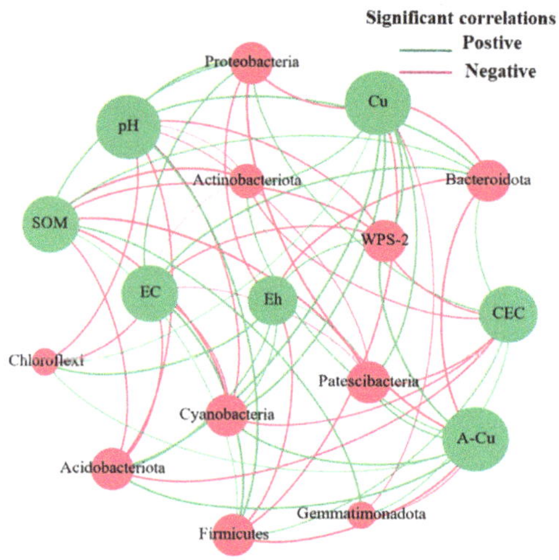

Figure 6. Co-occurrence network of soil physicochemical properties and dominant bacterial phyla in soil. Note: the size of the circles represents the number of connections in the correlation lines, and the thickness of the lines indicates the strength of the correlation, with thicker lines corresponding to larger absolute correlation coefficients and vice versa.

A thermal chart was constructed to appraise the intricate correlation existing between soil physicochemical attributes and the relative abundance of bacterial taxa, specifically the preeminent genera (Figure 7). The findings delineate a conspicuous positive correlation between *Phenylobacterium*, *Parasegetibacter*, *Micromonospora*, and soil EC, juxtaposed with a discernible negative correlation with the concentration of effective Cu within the soil matrix. Notably, robust positive correlation surfaces between soil SOM and the genera *Lysobacter* and *norank_f__norank_o__SBR1031*. Additionally, the genus Sphingomonas manifests a marked positive correlation with soil pH while concurrently exhibiting a markedly negative correlation with soil Eh and the aggregate concentration of total Cu. *Micromonospora* has been validated for its efficacy in fostering the growth of vegetation in soils laden with heavy metal contaminants [50]. The genera *Lysobacter* and *Sphingomonas*, owing to their pronounced adaptability to soil environments, exhibit notable potential in the remediation of soils subjected to heavy metal stress [51]. Ergo, the application of vermicompost and steel slag as amendments for the rectification of acidic soils hosting copper sulfide, begets alterations in the physicochemical milieu of the soil. Subsequently, this instigates a reshaping of the microbial community structure, ultimately culminating in an augmentation of soil quality.

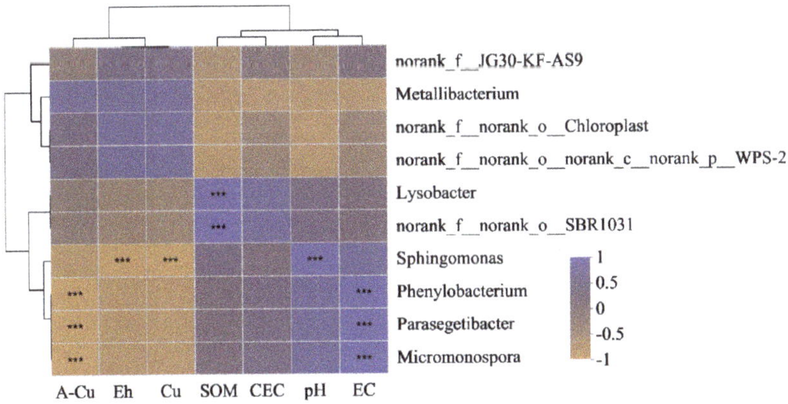

Figure 7. Heatmap of the correlation between soil physicochemical properties and the dominant bacterial phyla in soil. Note: *** $p \leq 0.001$.

4. Conclusions

This research endeavors to ameliorate acidic soils contaminated with copper sulfide via the application of a composite of *L. perenne* in conjunction with vermicompost and steel slag, either administered singularly or in tandem. The findings posit that *L. perenne* demonstrates notable resilience in adverse environments, rendering it a prospective phytoremediation candidate for soils afflicted by both acidity and copper pollutants. The solitary or combined application of vermicompost and steel slag substantively elevates soil parameters such as pH, electrical conductivity (EC), soil organic matter (SOM), and cation exchange capacity (CEC). This augmentation engenders an amplification in the abundance of bacterial communities concomitant with a diminution in soil redox potential (Eh), total copper content, and available copper content. It is noteworthy that the application of steel slag exhibits superior efficacy in immobilizing copper within the soil matrix, while vermicompost conduces to an augmentation in the diversity of soil bacterial communities. The ensuing alterations in soil physicochemical properties further precipitate a transformative impact on the composition of bacterial communities, and the joint application of steel slag and vermicompost serves to both enrich and stabilize the structure of soil bacterial communities. Thus, the synergistic application of vermicompost and steel slag in soils contaminated with heavy metals emerges as an environmentally conscientious and economically viable strategy for remediation.

Author Contributions: Conceptualization, J.X.; formal analysis, S.W.; data curation, H.Q.; writing—original draft preparation, X.W.; writing—review and editing, X.W., J.X., M.H., H.Q., and S.W.; supervision, J.X. and M.H.; funding acquisition, J.X. and X.W. All authors have read and agreed to the published version of the manuscript.

Funding: This work was grateful to the Key R&D Program of Jiangxi Province, China (20212BBG73013), and Jiangxi Province Graduate Innovation Special Fund Project, China (XY2022-S211).

Institutional Review Board Statement: Not applicable.

Informed Consent Statement: Not applicable.

Data Availability Statement: The datasets used and analyzed during the current study are available from the corresponding author upon reasonable request.

Acknowledgments: Thanks to Jiangxi University of Science and Technology for providing experimental facilities and instruments.

Conflicts of Interest: The authors declare no conflicts of interest.

References

1. Zhao, P.; Chen, J.; Liu, T.; Wang, Q.; Wu, Z.; Liang, S. Heavy metal pollution and risk assessment of tailings in one low-grade copper sulfide mine. *Front. Environ. Sci.* **2023**, *11*, 1132268. [CrossRef]
2. De-Conti, L.; Ceretta, C.A.; Melo, G.W.B.; Tiecher, T.L.; Silva, L.O.S.; Garlet, L.P.; Mimmo, T.; Cesco, S.; Brunetto, G. Intercropping of young grapevines with native grasses for phytoremediation of Cu-contaminated soils. *Chemosphere* **2019**, *216*, 147–156. [CrossRef]
3. Xu, D.; Fu, R.; Wang, J.; Shi, Y.; Guo, X. Chemical stabilization remediation for heavy metals in contaminated soils on the latest decade: Available stabilizing materials and associated evaluation methods—A critical review. *J. Clean. Prod.* **2021**, *321*, 128730. [CrossRef]
4. Tan, G.; Yu, H. Rethinking biochar: Black gold or not? *Nat. Rev. Mater.* **2023**, *9*, 4–5. [CrossRef]
5. Xu, D.; Carswell, A.; Zhu, Q.; Zhang, F.; De-Vries, W. Modelling long-term impacts of fertilization and liming on soil acidification at Rothamsted experimental station. *Sci. Total Environ.* **2020**, *713*, 136249. [CrossRef]
6. Swati, A.; Hait, S. Fate and bioavailability of heavy metals during vermicomposting of various organic wastes—A review. *Process Saf. Environ.* **2017**, *109*, 30–45. [CrossRef]
7. He, H.; Tam, N.F.Y.; Yao, A.; Qiu, R.; Li, W.; Ye, Z. Growth and Cd uptake by rice (*Oryza sativa*) in acidic and Cd-contaminated paddy soils amended with steel slag. *Chemosphere* **2017**, *189*, 247–254. [CrossRef]
8. Cutillas-Barreiro, L.; Fernandez-Calvino, D.; Nunez-Delgado, A.; Fernandez-Sanjurjo, M.J.; Alvarez-Rodriguez, E.; Carlos-Novoa-Munoz, J.; Arias-Estevez, M. Pine Bark Amendment to Promote Sustainability in Cu-Polluted Acid Soils: Effects on *Lolium perenne* Growth and Cu Uptake. *Water Air Soil Pollut.* **2017**, *228*, 260. [CrossRef]
9. Luo, L.; Xie, L.; Jin, D.; Mi, B.; Wang, D.; Li, X.; Dai, X.; Zou, X.; Zhang, Z.; Ma, Y. Bacterial community response to cadmium contamination of agricultural paddy soil. *Appl. Soil Ecol.* **2019**, *139*, 100–106. [CrossRef]
10. Suvendu, D.; Gwon, H.S.; Muhammad, I.K.; Jeong, S.K.; Kim, P.J. Steel slag amendment impacts on soil microbial communities and activities of rice (*Oryza sativa* L.). *Sci. Rep.* **2020**, *10*, 6746. [CrossRef]
11. Chaoui, H.I.; Zibilske, L.M.; Ohno, T. Effects of earthworm casts and compost on soil microbial activity and plant nutrient availability. *Soil Biol. Biochem.* **2003**, *35*, 295–302. [CrossRef]
12. Lu, R. *Analytical Methods for Soil Agrochemistry*; Chinese Agricultural Science and Technology Publishing House: Beijing, China, 2000; pp. 205–226.
13. Bao, S. *Soil and Agrochemistry Analysis*; China Agricultural Press: Beijing, China, 2000; pp. 25–38.
14. Kahr, G.; Madsen, F.T. Determination of the cation exchange capacity and the surface area of bentonite, illite and kaolinite by methylene blue adsorption. *Appl. Clay Sci.* **1995**, *9*, 327–336. [CrossRef]
15. Cottes, J.; Saquet, A.; Palayret, L.; Husson, O.; Beghin, R.; Allen, D.; Guiresse, M. Effects of soil redox potential (Eh) and pH on growth of sunflower and wheat. *Arch. Agron. Soil Sci.* **2020**, *66*, 473–487. [CrossRef]
16. Chen, S.; Zhou, Y.; Chen, Y.; Gu, J. Fastp: An ultra-fast all-in-one FASTQ preprocessor. *Bioinformatics* **2018**, *34*, 884–890. [CrossRef] [PubMed]
17. Isaac, R.; Siddiqui, S.; Aldosari, O.F.; Uddin, M.K. Magnetic biochar derived from Juglans regia for the adsorption of Cu^{2+} and Ni^{2+}: Characterization, Modelling, Optimization, and Cost analysis. *J. Saudi Chem. Soc.* **2023**, *27*, 101749. [CrossRef]
18. Anamika, S.; Savita, S.; Sonali, S.; Nitika, S.; Satveer, S.; Rahil, D.; Adarsh, P.V.; Avinash, K.N. Bio-conversion of Jamun leaf litter and kitchen waste into vermicompost: Implications for *Withania somnifera* (L.) Dunal in vitro conservation. *Biomass Conv. Bioref.* **2023**, 1–13. [CrossRef]
19. Ashok-Kumar, K.; Subalakshmi, R.; Jayanthi, M.; Abirami, G.; Vijayan, D.S.; Venkatesa-Prabhu, S.; Baskaran, L. Production and characterization of enriched vermicompost from banana leaf biomass waste activated by biochar integration. *Environ. Res.* **2023**, *219*, 115090. [CrossRef] [PubMed]
20. Soobhany, N.; Gunasee, S.; Rago, Y.P.; Joyram, H.; Raghoo, P.; Mohee, R.; Garg, V.K. Spectroscopic, thermogravimetric and structural characterization analyses for comparing Municipal Solid Waste composts and vermicomposts stability and maturity. *Bioresour. Technol.* **2017**, *236*, 11–19. [CrossRef]
21. Da-Silva, M.D.; Schnorr, C.; Lütke, S.F.; Silva, L.F.O.; Manera, C.; Perondi, D.; Godinho, M.; Collazzo, C.C.; Dotto, G.L. Citrus fruit residues as alternative precursors to developing H_2O and CO_2 activated carbons and its application for Cu(II) adsorption. *Environ. Sci. Pollut. Res.* **2023**, *30*, 63661–63677. [CrossRef]
22. Zhang, P.; Zhang, X.; Yuan, X.; Xie, R.; Han, L. Characteristics, adsorption behaviors, Cu(II) adsorption mechanisms by cow manure biochar derived at various pyrolysis temperatures. *Bioresour. Technol.* **2021**, *331*, 125013. [CrossRef]
23. Manchisi, J.; Matinde, E.; Rowson, N.A.; Simmons, M.J.H.; Simate, G.S.; Ndlovu, S.; Mwewa, B. Ironmaking and Steelmaking Slags as Sustainable Adsorbents for Industrial Effluents and Wastewater Treatment: A Critical Review of Properties, Performance, Challenges and Opportunities. *Sustainability* **2020**, *12*, 2118. [CrossRef]
24. Yang, M.; Lu, C.; Quan, X.; Cao, D. Mechanism of Acid Mine Drainage Remediation with Steel Slag: A Review. *ACS Omega* **2021**, *6*, 30205–30213. [CrossRef]

25. He, H.; Tam, N.F.Y.; Yao, A.; Qiu, R.; Li, W.; Ye, Z. Effects of alkaline and bioorganic amendments on cadmium, lead, zinc, and nutrient accumulation in brown rice and grain yield in acidic paddy fields contaminated with a mixture of heavy metals. *Environ. Sci. Pollut. Res.* **2016**, *23*, 23551–23560. [CrossRef] [PubMed]
26. Wang, L.; Fu, P.; Ma, Y.; Zhang, X.; Zhang, Y.; Yang, X. Steel slag as a cost-effective adsorbent for synergic removal of collectors, Cu(II) and Pb(II) ions from flotation wastewaters. *Miner. Eng.* **2022**, *183*, 107593. [CrossRef]
27. O'Connor, J.; Nguyen, T.B.T.; Honeyands, T.; Monaghan, B.; O'Dea, D.; Rinklebe, J.; Vinu, A.; Hoang, S.A.; Singh, G.; Kirkham, M.B.; et al. Production, characterisation, utilisation, and beneficial soil application of steel slag: A review. *J. Hazard. Mater.* **2021**, *419*, 126478. [CrossRef] [PubMed]
28. Nur-Sa'adah, A.H.; Rosazlin, A.; Karsani, S.A.; Osman, N.; Qurban, A.P.; Ishak, C.F. Influence of Soil Amendments on the Growth and Yield of Rice in Acidic Soil. *Agronomy* **2018**, *8*, 165. [CrossRef]
29. Chen, G.; Shah, K.J.; Shi, L.; Chiang, P.; You, Z. Red soil amelioration and heavy metal immobilization by a multi-element mineral amendment: Performance and mechanisms. *Environ. Pollut.* **2019**, *254*, 112964. [CrossRef] [PubMed]
30. Xu, M.; Xia, H.; Wu, J.; Yang, G.; Zhang, X.; Peng, H.; Yu, X.; Li, L.; Xiao, H.; Qi, H. Shifts in the relative abundance of bacteria after wine-lees-derived biochar intervention in multi metal-contaminated paddy soil. *Sci. Total Environ.* **2017**, *599–600*, 1297–1307. [CrossRef]
31. Liang, S.; Jin, Y.; Liu, W.; Li, X.; Shen, S.; Ding, L. Feasibility of Pb phytoextraction using nano-materials assisted ryegrass: Results of a one-year field-scale experiment. *J. Environ. Manag.* **2017**, *190*, 170–175. [CrossRef]
32. Jing, F.; Chen, C.; Chen, X.; Liu, W.; Wen, X.; Hu, S. Cadmium transport in red paddy soils amended with wheat straw biochar. *Environ. Monit. Assess.* **2021**, *193*, 381. [CrossRef]
33. Wang, F.; Shen, X.; Wu, Y.; Wang, Y.; Zhang, H.; Ding, Y.; Zhu, W. Evaluation of the effectiveness of amendments derived from vermicompost combined with modified shell powder on Cd immobilization in Cd-contaminated soil by multiscale experiments. *Ecotoxicol. Environ. Saf.* **2023**, *262*, 115166. [CrossRef]
34. Ning, D.; Liang, Y.; Song, A.; Duan, A.; Liu, Z. In situ stabilization of heavy metals in multiple-metal contaminated paddy soil using different steel slag-based silicon fertilizer. *Environ. Sci. Pollut. Res.* **2016**, *23*, 23638–23647. [CrossRef]
35. Liu, H.; Zhang, T.; Tong, Y.; Zhu, Q.; Huang, D.; Zeng, X. Effect of humic and calcareous substance amendments on the availability of cadmium in paddy soil and its accumulation in rice. *Ecotoxicol. Environ. Saf.* **2022**, *231*, 113186. [CrossRef] [PubMed]
36. Jones, S.; Bardos, R.P.; Kidd, P.S.; Mench, M.; De-Leij, F.; Hutchings, T.; Cundy, A.; Joyce, C.; Soja, G.; Friesl-Hanl, W.; et al. Biochar and compost amendments enhance copper immobilisation and support plant growth in contaminated soils. *J. Environ. Manag.* **2016**, *171*, 101–112. [CrossRef]
37. Mathieu, S.; Steve, P.; Fernando, P.; Frédéric, P.; Jacques, M.; Noureddine, M.; Olivier, F. Aided-phytostabilization of steel slag dumps: The key-role of pH adjustment in decreasing chromium toxicity and improving manganese, phosphorus and zinc phytoavailability. *J. Hazard. Mater.* **2021**, *405*, 124225. [CrossRef]
38. Nadia, K.; Rafael, C.; Eduardo, M.; Nicholas, W.L.; Luke, B. Efficiency of green waste compost and biochar soil amendments for reducing lead and copper mobility and uptake to ryegrass. *J. Hazard. Mater.* **2011**, *191*, 41–48. [CrossRef]
39. Sarathchandra, S.S.; Rengel, Z.; Solaiman, Z.M. Remediation of heavy metal-contaminated iron ore tailings by applying compost and growing perennial ryegrass (*Lolium perenne* L.). *Chemosphere* **2022**, *288*, 132573. [CrossRef] [PubMed]
40. Jiang, B.; Adebayo, A.; Jia, J.; Xing, Y.; Deng, S.; Guo, L.; Liang, Y.; Zhang, D. Impacts of heavy metals and soil properties at a Nigerian e-waste site on soil microbial community. *J. Hazard. Mater.* **2019**, *362*, 187–195. [CrossRef]
41. Drzewiecka, D. Significance and Roles of Proteus spp. Bacteria in Natural Environments. *Microb. Ecol. Int. J.* **2016**, *72*, 741–758. [CrossRef]
42. Altimira, F.; Yáñez, C.; Bravo, G.; González, M.; Rojas, L.A.; Seeger, M. Characterization of copper-resistant bacteria and bacterial communities from copper-polluted agricultural soils of central Chile. *BMC Microbiol.* **2012**, *12*, 193. [CrossRef]
43. Sadaf, K.; Anirban, B.; Iqbal, A.; Sayyed, R.Z.; Hesham, A.E.; Daniel, J.D.; Ni, L.S. Recent Understanding of Soil Acidobacteria and Their Ecological Significance: A Critical Review. *Front. Microbiol.* **2020**, *11*, 580024. [CrossRef]
44. Andriy, S.; Gareth, M.J.; Jessica, J.; Robert, M.B.; Isaac, B.; Cassandra, C.; Emiley, A.E.; Natalia, I.; Rex, R.M.; Stephen, E.G.; et al. Ecological and genomic analyses of candidate phylum WPS-2 bacteria in an unvegetated soil. *Environ. Microbiol.* **2020**, *22*, 3143–3157. [CrossRef]
45. Wang, F.; Zhang, W.; Miao, L.; Ji, T.; Wang, Y.; Zhang, H.; Ding, Y.; Zhu, W. The effects of vermicompost and shell powder addition on Cd bioavailability, enzyme activity and bacterial community in Cd-contaminated soil: A field study. *Ecotoxicol. Environ. Saf.* **2021**, *215*, 112163. [CrossRef]
46. Cui, J.; Yang, B.; Zhang, M.; Song, D.; Xu, X.; Ai, C.; Liang, G.; Zhou, W. Investigating the effects of organic amendments on soil microbial composition and its linkage to soil organic carbon: A global meta-analysis. *Sci. Total Environ.* **2023**, *894*, 164899. [CrossRef]
47. Hu, H.; Shao, T.; Gao, X.; Long, X.; Rengel, Z. Effects of planting quinoa on soil properties and microbiome in saline soil. *Land Degrad. Dev.* **2022**, *33*, 2689–2698. [CrossRef]
48. Hahm, M.; Son, J.; Kim, B.; Ghim, S. Comparative study of rhizobacterial communities in pepper greenhouses and examination of the effects of salt accumulation under different cropping systems. *Arch. Microbiol.* **2017**, *199*, 303–315. [CrossRef] [PubMed]
49. Mujakic, I.; Piwosz, K.; Koblízek, M. Phylum Gemmatimonadota and Its Role in the Environment. *Microorganisms* **2022**, *10*, 151. [CrossRef] [PubMed]

50. Ortúzar, M.; Trujillo, M.E.; Román-Ponce, B.; Carro, L. *Micromonospora metallophores*: A plant growth promotion trait useful for bacterial-assisted phytoremediation? *Sci. Total Environ.* **2020**, *739*, 139850. [CrossRef] [PubMed]
51. Hu, X.; Wang, J.; Lv, Y.; Liu, X.; Zhong, J.; Cui, X.; Zhang, M.; Ma, D.; Yan, X.; Zhu, X. Effects of Heavy Metals/Metalloids and Soil Properties on Microbial Communities in Farmland in the Vicinity of a Metals Smelter. *Front. Microbiol.* **2021**, *12*, 707786. [CrossRef] [PubMed]

MDPI AG
Grosspeteranlage 5
4052 Basel
Switzerland
Tel.: +41 61 683 77 34

Applied Sciences Editorial Office
E-mail: applsci@mdpi.com
www.mdpi.com/journal/applsci

www.ingramcontent.com/pod-product-compliance
Lightning Source LLC
Chambersburg PA
CBHW062057220926
51850CB00010B/2802
9783725849079